Manufacturing Operations Management

Manufacturing Operations Management

Dr. Sanjay Sharma
National Institute of Industrial Engineering (NITIE), Mumbai, India

Taylor & Francis
Taylor & Francis Group
Boca Raton London New York
CRC is an imprint of the Taylor & Francis Group, an informa business

Ane Books Pvt. Ltd.

Manufacturing Operations Management

Dr. Sanjay Sharma

Published by

Ane Books Pvt. Ltd.

4821 Parwana Bhawan, 1st Floor
24 Ansari Road, Darya Ganj, New Delhi - 110 002, India
Tel: +91 (011) 2327 6843-44, 2324 6385
Fax: +91 (011) 2327 6863
e-mail: anebooks@vsnl.net
Website: www.anebooks.com

For

CRC Press

Taylor & Francis Group
6000 Broken Sound Parkway, NW, Suite 300
Boca Raton, FL 33487 U.S.A.
Tel : 561 998 2541
Fax : 561 997 7249 or 561 998 2559
Web : www.taylorandfrancis.com

For distribution in rest of the world other than the Indian sub-continent

ISBN : 978-14-8225-788-5

British Library Cataloguing in Publication Data
A catalogue record for this book is available from the British Library

Printed at: Thomson Press, India

Preface

Manufacturing operations management is an important subject for engineering/technical as well as management stream. It includes activities related to manufacturing and associated services. The objective of the present book is to introduce the manufacturing and associated operational management techniques to the readers. The chapter on "Facilities Planning" incorporates certain computerized techniques in addition to providing relevant input information concerned with different types of layout.

With reference to the chapter on "Financial Management", emphasis is on making it more analysis-oriented. The professional engineers will also find it as a good reference on financial analysis and planning. They may also find the present work (which includes demand forecasting) helpful in refreshing their knowledge. Finally, considering the extreme business/industrial importance of the subject, I hope that the book will be highly useful for students and practitioners.

—Sanjay Sharma

Contents

1

Introduction

Whenever we visit an industrial organization, the following questions come to our mind:

(*a*) Why is this type of industry situated at this place only?

(*b*) Why are various departments and facilities located in a particular order?

(*c*) How do they predict the customer demands and plan their production/service accordingly?

(*d*) What kinds of techniques are adopted in order to improve the management and operational efficiencies with an objective of maximizing the profit or minimizing the costs?

(*e*) How do they finance various projects and other routine activities, as survival of the organization becomes difficult without effective financial management?

The subsequent chapters make an attempt to answer these relevant questions.

1.1 Product Life Cycle

Finished products offered by the industries to various customers have a certain life cycle. A product life cycle is shown in Fig. 1.1.

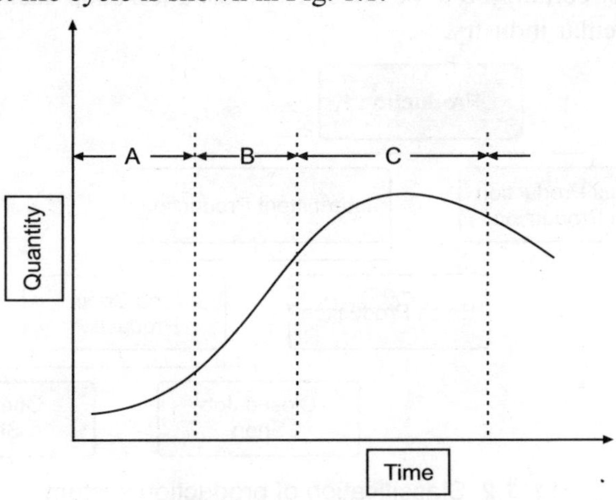

Fig. 1.1 Product life cycle

This is a relationship between quantity of production/sales volume and time. It is divided into the following four zones:

A. Introduction

In this zone, the product is introduced to the customer. Initially, sales volume is less, and then it increases at a slow rate.

B. Growth

In this zone, the demand of a product grows at a fast rate. It is an indication that more number of customers have started liking the product.

C. Maturity

In the maturity zone, the product becomes mature. Demand of the product is at its peak and almost constant.

D. Decline

Demand of the product declines once the maturity period is over. This may be due to the following reasons:

I. Product becomes obsolete.

II. Competitors have started playing an effective role and other models of the product have appeared in the market.

Before a decline in the demand of the product occurs, diversification is necessary. An analogy may be "if the road ahead is difficult, then there is diversion". In order to avoid decline, design of the product may be modified. Utility as well as reliability of the product may be enhanced. Additional feature are incorporated in the product to create the demand.

To meet the customer demands, different types of production system may be adopted. Fig 1.2 shows a broad classification of production.

In continuous production or mass production, the demand is more than the production rate. The production rate is more than the demand rate in intermittent production.

Job shops are of two types–closed job shop and open job shop. An example of the open job shop is a common welding shop or a machine shop, where anyone can approach for certain job to be done. The closed job shop usually caters to the needs of a particular industry.

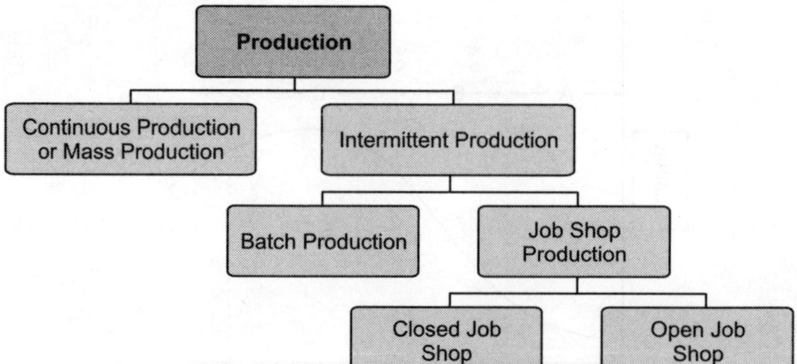

Fig. 1.2 Classification of production system

For example, a large water pump manufacturing industry may have its own job shop in the same premises in addition to various departments such as casting, machining, etc. Forecasting of demands is easier for closed job shops as compared to open job shops. In job shop production, variety of products is very high but quantity of production for each variety is very less.

On the other hand, in mass production, quantity of production is high; however, only a few variety of products, sometime even a single standard products is manufactured. Therefore, characteristic features such as variety and quality of products, mass production and job shop production are at two extremes. Batch production is at intermediate level. Refer to Fig 1.1, during product life cycle, it seems convenient to adopt (*a*) job shop production in the introduction zone, (*b*) batch production in the growth region, and (*c*) mass production in the maturity zone.

Production rate is more than the demand rate in batch production as shown in Fig. 1.3. During cycle time, T, production activity takes place for a portion of time and the remaining of the cycle time is idle. During the idle time, the facility is free for other tasks, if available.

1.2 Operations Management

The concepts discussed before are not applicable to a manufacturing situation but for various kinds of services. That is why the name "operations management", which is a broader discipline encompassing all kinds of operations whether producing tangible goods or providing services. For example, in IT industry, no tangible goods are produced but software programs are generated for different industrial or business functions. Product life cycle is very much relevant for IT services.

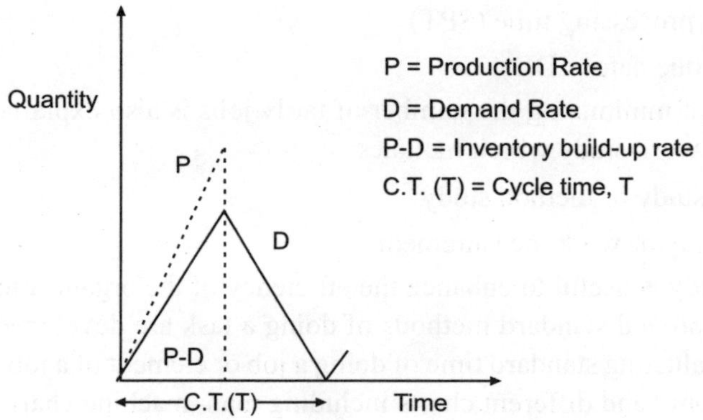

Fig. 1.3 Batch production

Demand forecasting has an important role to play in designing a manufacturing system or providing services to customers. A strategic decision is taken after the study of demand patterns.

Next chapter deals with facilities planning. This is related to arrangement of facilities or departments in an organisation. A process or functional layout is suitable for intermittent productions and a product or line layout is suitable for continuous production. Functional layout is useful for many applications such as hospitals but state of the patient changes by utilizing resources such as medicines, doctors etc. as compared to movements of goods in a factory, movement of persons in the hospitals becomes a major criterion in arranging the departments in process or functional type of layout. Minimizing the total handling cost is one of the main objectives in such type of layouts where line balancing is the important objective in line or product layout. Departments are also placed adjacent to each other on the basis of their closeness rating. Some computerized techniques are discussed for layout planning in chapter 2.

Chapter 3 discusses forecasting of demands. Forecasting method may be qualitative or quantitative in nature. Qualitative method may be adopted when previous data pertaining to demand are either not available or not sufficient. Quantitative techniques are useful when a rich record of previous data is available. However, a combination of both the techniques may prove to be more beneficial.

Production planning and control, *i.e.* PPC ,is described in chapter 4. Production planning at higher level is based on demand forecasts, whereas at the shop floor, following PPC techniques are adopted:

I. Routing

II. Scheduling

III. Despatching

IV. Follow-up

Different jobs queue up before a single machine in a job shop. Sequencing of jobs on the basis of following rules is discussed:

I. Shortest processing time (SPT)

II. Earliest due date (EDD)

Method of minimizing the number of tardy jobs is also explained. Chapter 5 is devoted to work study which includes:

I. Motion study or method study

II. Time study or work measurement

Work study is useful to enhance the efficiency of the organisation. In method study, the improved standard methods of doing a task are developed. Time study deals with evaluating standard time of doing a job or element of a job. Principles of motion economy and different charts including man–machine chart, flow process chart etc. are also explained.

In any industrial organization, following types of inventory are present:

I. Raw material inventory

II. Finished product inventory

III. Work-in-process inventory

Inventory management is described in chapter 6. Conflicting cost such as ordering cost and inventory caring costs are balanced in determining the economic order quantity (EOQ). EOQ is derived by minimizing the total cost. Following concepts are also discussed:

I. Lead time

II. Safety stock

III. Reorder points.

Chapter 7 deals with inspection and quality control. Inspection is detection of defects, while quality control attempts prevention of defects. For variables, $\overline{X}$- and R-charts are illustrated with an example. For attributes, P- and C-charts are explained with numerical examples. Acceptance sampling along with producers and consumers risk is also discussed.

Projects are the tasks which will be done at the same place perhaps once in several years. Examples may include construction of bridge, R&D project, installation of plant and machinery, etc. Project management is discussed in chapter 8. Network methods for project planning have been categorised as:

I. Critical path method (CPM)

II. Program evaluation and review technique (PERT)

The CPM deals with activities having deterministic time estimate. The PERT deals with probabilistic situation in which three time estimates are available for such activity. Sometimes, project duration needs to be shortened. This is done by crashing the activities on critical path. Crashing of the project network is explained with the help of a numerical example.

Chapter 9 deals with the financial management. Its significance is shown considering different problems which are often encountered in an organization. Financial analysis of proposals may follow the process:

I. Ranking of proposals by quantitative/qualitative approach

II. Evaluation of capital budgeting proposals by payback method/present value approaches

III. Proposal is accepted or rejected or decision is kept pending

Time value of money is explained with suitable examples. Securities shares and bonds and valuation of securities is discussed along with measurement of risk. Cost of capital and optimal capital structure are included. Capital budgeting deals with procurement decisions of fixed assets. Following methods are illustrated:

I. Payback period

II. Internal rate of return approach

III. Net present value method

❑❑❑

Inventory management is described in chapter 6. Conflicting cost such as ordering cost and inventory carrying costs are balanced in determining the economic order quantity (EOQ). EOQ is derived by minimizing the total cost. Following concepts are also discussed:

I. Lead time

II. Safety stock

III. Reorder point

Chapter 7 deals with inspection and quality control. Inspection is detection of defects, whereas quality control attempts prevention of defects. For variables, X-bar and R-charts are illustrated with an example. For attributes, P- and C-charts are explained with numerical examples. Acceptance sampling plans with producers and consumers risk is also discussed.

Projects are the tasks which will be done at the same place, perhaps once in several years. Examples may include construction of bridge, R&D project, installation of plant and machinery, etc. Project management is discussed in chapter 8. Network methods for project planning have been categorized as:

I. Critical path method (CPM)

II. Program evaluation and review technique (PERT)

The CPM deals with activities having deterministic time estimate. The PERT deals with probabilistic situation in which three time estimates are available for such activity. Sometimes, project duration needs to be shortened. This is done by crashing the activities on critical path. Crashing of the project network is explained with the help of a numerical example.

Chapter 9 deals with the financial management. Its significance is shown considering different problems which are often encountered in an organization. Financial analysis of proposals may follow the process:

I. Ranking of proposals by quantitative/qualitative approach

II. Evaluation of capital budgeting proposals by payback method/present value approaches

III. Proposal is accepted or rejected or decision is kept pending

Time value of money is explained with suitable examples. Securities, shares and bonds and valuation of securities is discussed along with measurement of risk. Cost of capital and optimal capital structure are included. Capital budgeting deals with investment decisions of fixed assets. Following methods are illustrated:

I. [illegible] period

II. Internal rate of return approach

III. Net present value method

[illegible]

2

Facilities Planning

The objective of any entrepreneur is to locate the industrial or business organization at such a place where it is convenient to run the operations and the total costs are minimized. Once a decision is taken about the plant location, the next problem is how to arrange various facilities, *i.e.* equipment, service centres, departments, etc. This chapter deals with the plant location and layout planning. The discussions include factors affecting plant localization, selection of site, material flow pattern, different types of plant layout and the computerized layout planning procedures.

2.1 Plant Location

The ideal location for any industrial organization is depicted in Fig. 2.1.

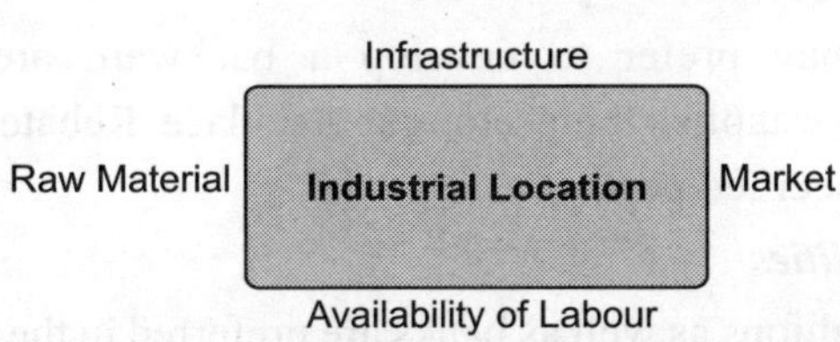

Fig. 2.1 Ideal location of an industrial organization

This is the situation in which raw materials as well as market for finished goods are near to the industry. Easy availability of labour and infrastructure including basic facilities such as power, water and transportation exist.

But in the real environment, ideal conditions do not always exist. The effort should be to avail maximum benefits. Numerous factors affecting plant localization are categorized as follows:

I. Significant Factors

1. ***Proximity to raw materials and market***

   If the source of raw materials is in the nearby area of industrial location, it is advantageous from the point of view of transportation cost. Similarly, if the location is closer to a market place from where the finished products are available to consumer, the cost of transportation can be kept to a minimum.

In case either the raw material or finished product (or both) is having a shelf life constraint, *i.e.* these can be spoiled in a specified time period (*e.g.* in food processing industry), proximity of raw materials/market becomes critical.

2. *Infrastructure*

Efficient transportation system along with the built-up roads and smooth availability of power and water are necessary for industrial or economic activities.

3. *Availability of Labour*

As long as the manpower is cheaper than automation and if it does not affect the quality of product, the availability of labour will be a major issue in some of the industrial activities.

II. Considerable Factors

1. *Availability of supplementary industries in proximity*

An industry requires many components some of which are purchased from outside sources. If the ancillary units manufacturing these components are in proximity of the main industry, then it is always preferred. This factor is becoming significant day-by-day because of the supply chain management and just-in-time inventory systems.

2. *Storage facilities*

The criterion gets utmost attention in some cases such as the agricultural industries.

3. *Government/taxation policy*

Government may prefer to develop a backward area by encouraging industrialists to establish their setups at that place. Rebate in the taxes may be offered for a specified period.

4. *Financial facilities*

Financial institutions as well as banks are preferred in the proximity due to an ease in the management of current assets/long-term loans.

III. Additional Factors

1. *Strategic factor*

All the industries should not be located at one place. Rather, these must be scattered throughout the nation. Due to a natural tragedy or war, one or few regions may be destroyed and if the industries are situated at several places in the country, then the whole economy will not be ruined.

2. *Size/price of the plot and its topography*

Size and shape of the available plot is to be taken into consideration for the layout design as will be explained later. Depending on the situation, price of the land also becomes important. Topography, *i.e.* ups and downs of the region is of concern from civil engineering point of view because of the levelling effort and cost involved.

2.2 Site Selection

There are basically two types of sites:

1. Urban area, and 2. Rural area.

1. ***Urban Area***

   Usually, the urban region has a good infrastructure and an easy availability of professionals and skilled labour. But, there is less scope for expansion of the industry. Price of the land is high. Pollution is a major concern.

2. ***Rural Area***

   The problem of pollution is less. Land is cheaper and more scope for expansion of the industry exists. Industrial relations are good. But, poor infrastructure and availability of professionals are limited as most of the professionals do not like to live in the rural area.

   After discussing the advantages and disadvantages of urban as well as rural areas, sub-urban area seems to be a compromise between urban and rural regions. Quite satisfactory infrastructure is available in sub-urban area and professionals also do not hesitate in living there.

   In Section 2.1 above, different factors were discussed some of which are difficult to quantify. However, an attempt should be made to compare various alternative sites by evaluating the total costs based on input items, the resources to convert them into finished products, selling expenditure and taxes.

The criteria are also classified as shown in Fig. 2.2.

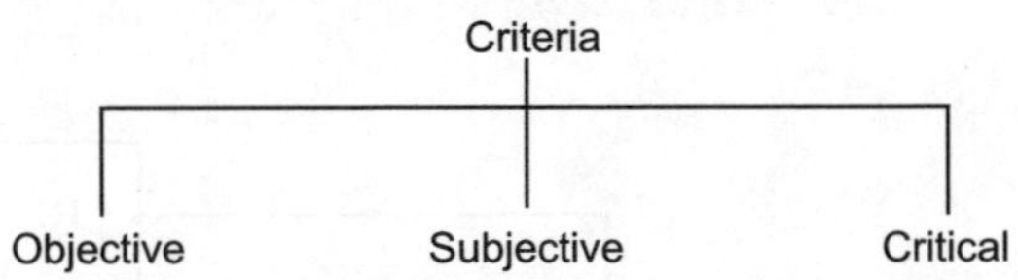

Fig. 2.2 Criteria affecting the selection of a location

- Objective factors are those which can be quantified in the money value
- Subjective factors are qualitative type of factors which can not be evaluated in money terms
- Critical factor is the criterion in the absence of which a particular site can never be considered

2.3 Material Flow Pattern

The raw material/input items move from one machine to another, till these become the finished items. The material flow lines refer to the flow pattern. There are five basic flow lines:

(*a*) I-Flow
(*b*) L-Flow
(*c*) S-Flow
(*d*) O-Flow
(*e*) U-Flow

The flow lines are according to the shape of the letter, *i.e.*, 'I', 'L', 'S', 'O', or 'U'

(*a*) I-Flow

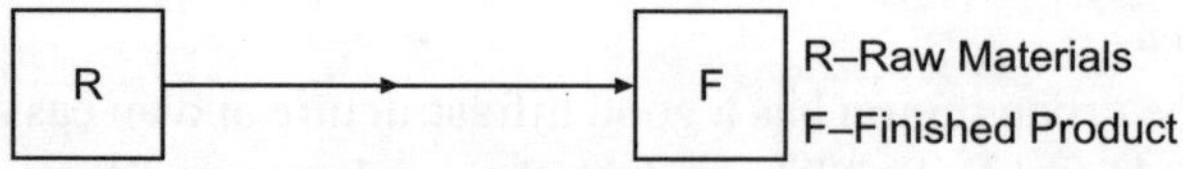

At location 'R', the raw material enters and at location 'F', the finished product is available for dispatch. The machines and facilities are arranged to the line joining 'R' and 'F' as per the requirement. The type of flow line is suitable for the available space, which is more in length and less in width.

(*b*) L-Flow

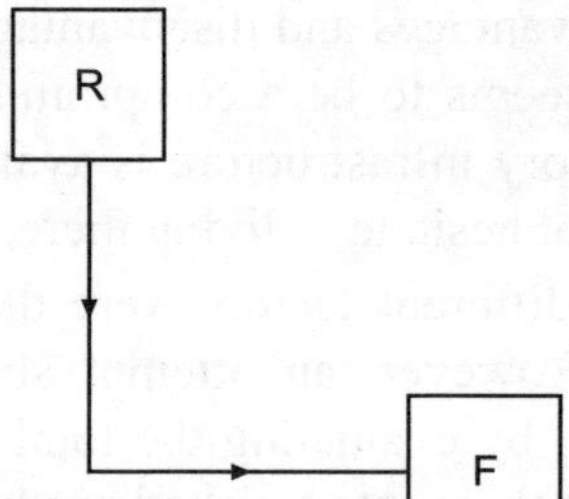

This is used to take an advantage of width of the plot also, if it is sufficiently wide.

(*c*) S-Flow

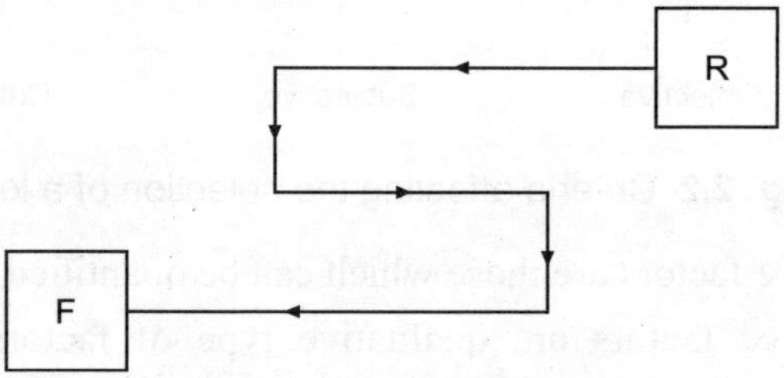

When the number of facilities are more and the available space suits, S-Flow may be adopted. It can also be imagined as a combination of U-Flow and L-Flow lines and some of the advantages of both the patterns can be availed.

(*d*) O-Flow

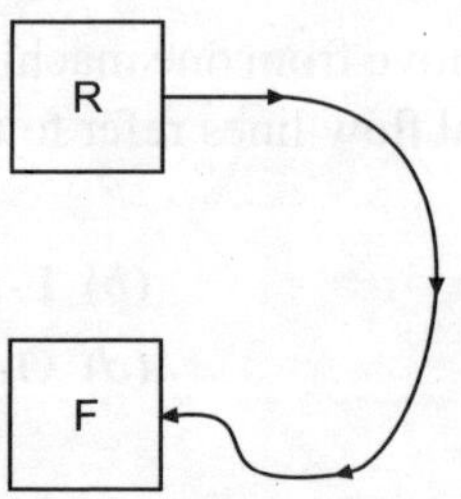

(*e*) U-Flow

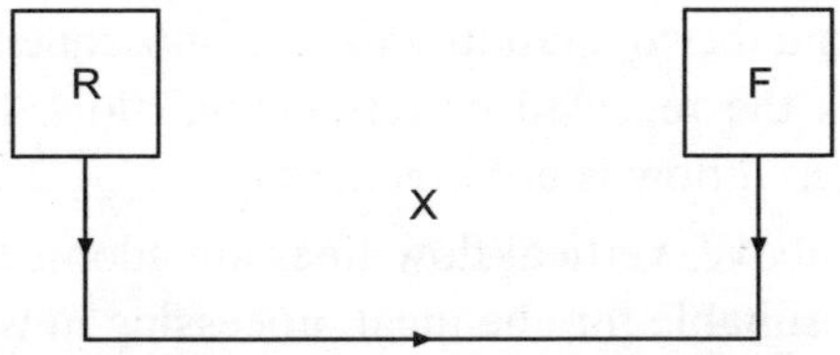

The facilities are arranged in U-shape. The engineers and supervisors need to travel less distance. For example, if a person stands at location ‘X’, it may be convenient to supervise the machines surrounding him.

Out of the five basic flow lines, in case of ‘O’ and ‘U’, the entry of raw material/input items, ‘R’, and dispatch of finished products, ‘F’, are on the same side of the plant. Vehicles will be coming in and going out for the transportation of raw materials and finished items respectively. Road is needed to be built up only on one side of the plant. But, at the same time, a heavy rush of transportation will be there. If the final product is smaller in size, the containers carrying them flow through the facilities. After dispatch of the final products, the empty containers are needed to be brought back to the input side of the plant. It becomes convenient in ‘O’- and ‘U’- flow patterns, because ‘R’ and ‘F’ are near to each other.

For I-, L- and S - slow patterns, the roads are required on both sides of the plant. But, a heavy rush of transportation is not there because the activities are divided on two sides of the layout for transportation of input and finished items. The handling cost for empty containers may be high.

In an industry, there may be a combination of two or more types of basic flow lines.

Sometimes, the repeated type of flow may also take place as shown in Fig. 2.3.

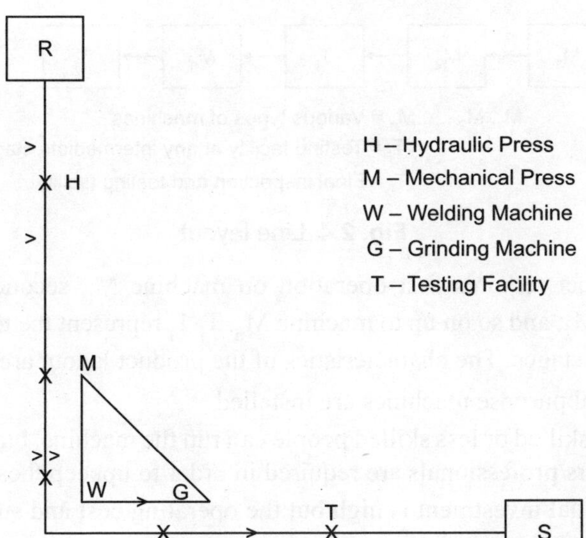

Fig. 2.3 Repeated flow

After the grinding operation is over, the part/product may again go to the mechanical press for another operation which is subsequent to the first grinding job. This is known as the repeated or retractional flow. In the absence of the repeated flow, the material flow is unidirectional.

In addition to the above, vertical flow lines are adopted to take the benefit of gravity. These may be suitable for chemical processing in which large vessels are placed on different floors one over the other. The vertical flow lines are also useful when horizontal space constraints exist.

2.4 Plant Layout

Plant layout refers to the arrangement of facilities, equipment and departments inside the industrial/business premises. The concept is useful for manufacturing as well as service type of organizations. The kinds of layout are as follows:

(*a*) Line or product layout
(*b*) Functional or process layout
(*c*) Fixed position layout
(*d*) Cellular manufacturing layout
(*e*) Hybrid layout

(*a*) ***Line or Product Layout***

Product layout is suitable for mass production where the volume of production is high and the variety of product is less, sometimes even a single standard product. The machines and testing facilities are arranged in the sequence of processes to be carried out and the tests at different stages. The layout is shown in Fig. 2.4.

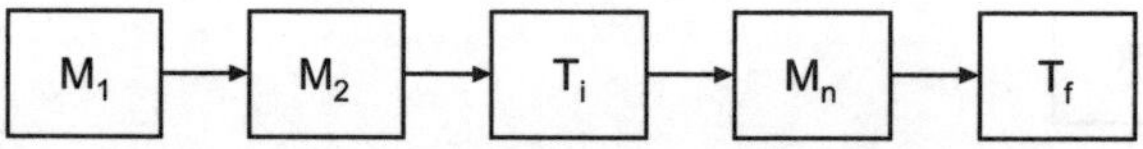

M_1, M_2 ......M_n = Various types of machines
T_i = Testing facility at any intermediate stage
T_f = Final inspection and testing facility

Fig. 2.4 Line layout

The product requires first operation on machine M_1, second operation on machine M_2, and so on up to machine M_n. T_i/T_f represent the testing facilities at various stages. The characteristics of the product layout are as follows:

1. Special purpose machines are installed.
2. Semi-skilled or less skilled people can run the machine, but highly skilled workers/professionals are required in order to upkeep these machines.
3. An initial investment is high but the operating cost and subsequently the production cost is low.
4. If one machine stops functioning and as it is the line layout, the entire production operation may come to a halt.

5. Conveyor system for material handling can be adopted for the smooth flow of material.
6. Work-in-progress inventory is less because the output of one machine immediately goes to the next machine if it is available to process.
7. All the machines are specialized for a particular design of the product. If a change in design is required, then it is difficult for the production system to respond to it. Therefore, flexibility in the system is not available.

For example, three machines exist in a product layout.

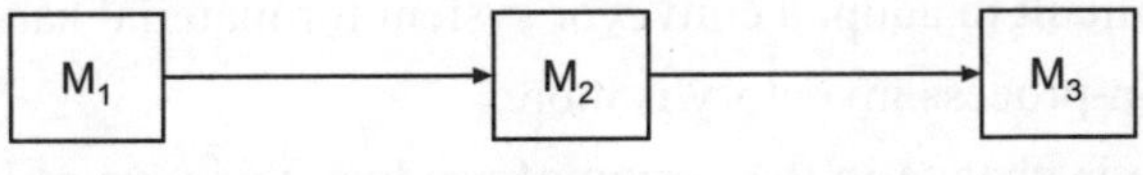

Production Capacity of M_1 and M_3 are 50 units per hour and the capacity of M_2 is 25 units per hour, then M_3 will be idle for half of the time. The solution may be to install two numbers of machine M_2 as shown below:

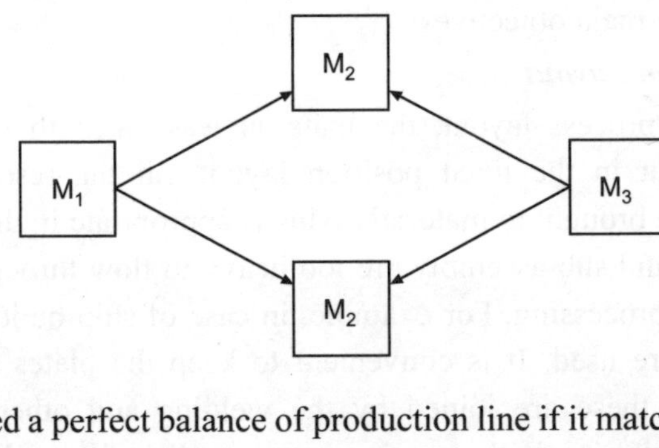

This is called a perfect balance of production line if it matches the production requirement. But in the real environment, when the number of machines are more and each has different capacity of production, it is difficult to balance the line completely. Hence, line balancing is a problem to be solved in this type of layout.

(*b*) ***Functional or Process Layout***

Process or functional layout is shown in Fig. 2.5 in which similar machines are grouped together at one location. However, it is not necessary for each product type to move through all types of machines. Depending on the requirements, any product may skip one or more departments.

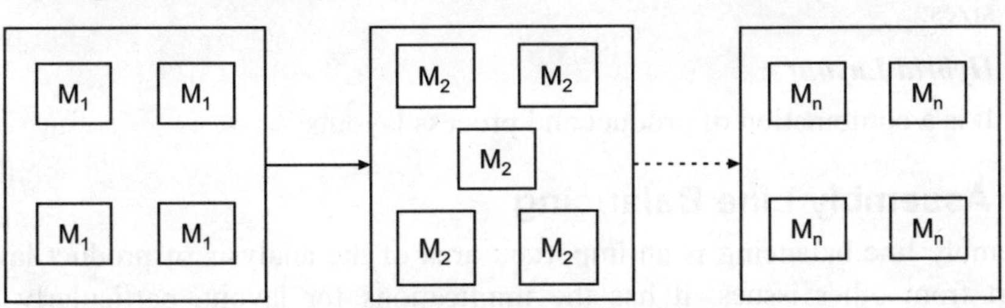

Fig. 2.5 Functional layout

Characteristics of this type of layout are as follows:

1. It is suitable for an intermittent production, *i.e.* the job shop or batch production.
2. General purpose machines are installed.
3. Highly skilled people are required to set up and run the machines.
4. An initial investment is less but the cost of production is high.
5. If there is a breakdown of one or two machines, the entire production process will not be stopped.
6. It is difficult to adopt a conveyor system for material handling.
7. Work-in-process inventory is more.
8. If there is change in the design of product, this type of layout is able to respond to it in a short time as the machines can be set up for a variety of products. Therefore, the system is flexible.
9. While designing the layout, minimizing the total material handling cost is one of the main objectives.

(*c*) ***Fixed Position Layout***

In a product/process layout, the material was taken to various facilities/ machines. But in the fixed position layout, all the resources, including machines, are brought to materials. This is appropriate in the situation when the material and sub-assembly are too heavy to flow through the stationary facilities for processing. For example, in case of ship building, heavy/thick steel plates are used. It is convenient to keep the plates and components together and these are joined by the welding and other processes. The welding machines and other equipments as well as the workers go to the site of material and a large ship is fabricated.

(*d*) ***Cellular Manufacturing Layout***

This is suitable for group technology programme. An existing job shop may select some of the parts manufactured by it which are suitable to be made in the cells. A cell acts like a product layout for such parts. A part family is a group of such parts which have similar operational requirements, *i.e.* they are processed on similar machines with identical setups. Further, the demand of such parts should be high so that they can be manufactured in moderate batch sizes.

(*e*) ***Hybrid Layout***

It is a combination of product and process layouts.

2.5 Assembly Line Balancing

Assembly line balancing is an important area of the analysis of product layout. Apart from other issues, it has the implications for layout, particularly with reference to the workstation size. A work station is a physical location where one or more tasks are performed. The objective is to assign tasks to workstations in such a way that each workstation completes the assigned tasks in less than or equal to the cycle time. A cycle time is the time between successive products coming off the end of an assembly line.

$$\text{Cycle Time (C.T.)} = \frac{\text{Production time per period}}{\text{Required production quantity per period}}$$

2.5.1 Largest Number of Following Tasks Rule

The requirements for implementing this rule are as follows:

1. Precedence diagram, which is a representation of relationship among tasks. For example, given:

Tasks	Tasks which Immediately Precede
A	–
B	A
C	A
D	B, C
E	D

The precedence diagram can be drawn as follows:

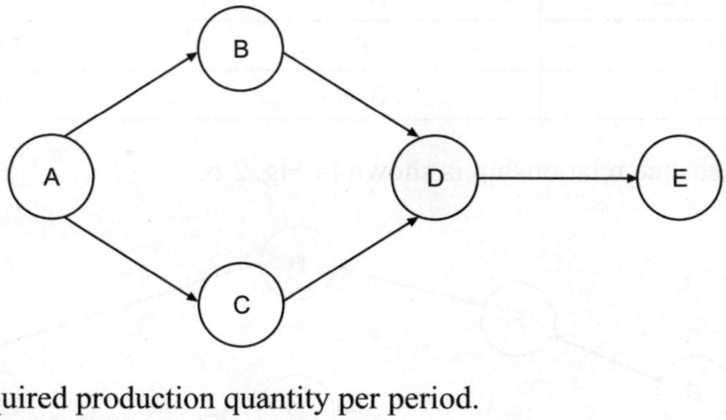

2. The required production quantity per period.
3. The time taken by each task, ti
4. As the name of the rule suggests, assignment of tasks is in the order of the largest number of following tasks. It will be helpful if the number of tasks which follow each task is known. Consider the discussed precedence diagram. First task is A and 4 number of tasks follow it, *viz*. B, C, D and E. Two tasks, *i.e.* D and E follow both B and C. Only one task E follows D and no task follows E because E, is the last task.

This can be summarized below as:

Tasks	Number of tasks which follow
A	4
B, C	2
D	1
E	0

Example 2.1

An assembly line involves tasks from A to I. 360 products are required per day. Production time is 8 hours per day. The time taken by each task and the tasks which immediately precede each task are:

Tasks	Tasks which Immediately Precede	Time to Complete Task (Seconds)
A	–	40
B	A	30
C	A	35
D	C	30
E	C	20
F	B	45
G	D, E	50
H	F	30
I	G, H	35

The precedence relationship is shown in Fig. 2.6.

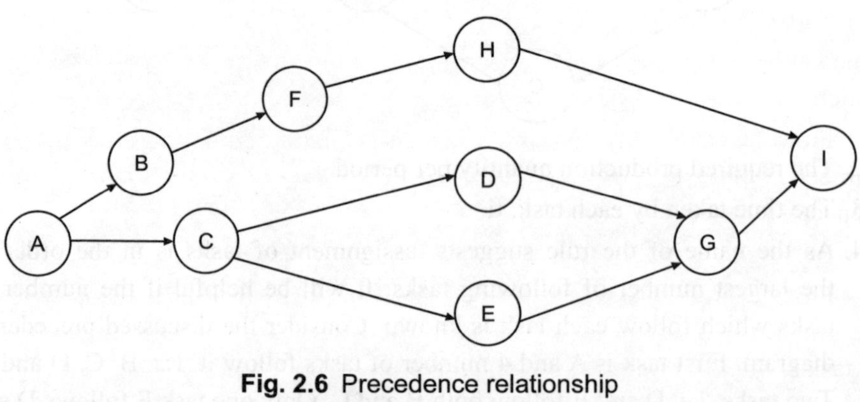

Fig. 2.6 Precedence relationship

As discussed before, it is necessary to know the total number of tasks, which follow each task because the assignment of task to each workstation will be in the order of the largest number of following tasks.

It is obtained below:

Tasks	A	B	C	D	E	F	G	H	I
Total number of tasks which follow it	8	3	4	2	2	2	1	1	0

Table 2.1 may be arranged in a decreasing order of the number of following tasks as shown below:

Table 2.1 Tasks in the order of largest number of following tasks first

Tasks	A	C	B	D	E	F	G	H	I
Total number of tasks which follow it	8	4	3	2	2	2	1	1	0

Before assigning the tasks to workstation, the cycle time needs to be computed because each workstation should not consume more than the cycle time in order to achieve the production target.

$$\text{Cycle Time} = \frac{\text{Production time per period}}{\text{Required production quantity per period}}$$

$$= \frac{8 \times 60 \times 60}{360} = 80 \text{ seconds}$$

Refer to Table 2.1. To begin with, the first task in the order is A, since it has the maximum number of the following tasks.

Workstation 1

The first task that is assigned to workstation 1 is A which takes 40 seconds.

Time which is yet to be assigned = 80 – 40

= 40 seconds.

Set of tasks which may be assigned at this stage considering precedence diagram (Fig. 2.6) only, *i.e.*, without checking the feasibility, S = [B, C].

Task time of B and C are 30 and 35 seconds respectively which is less than the time yet to be assigned, *i.e.* 40 seconds as calculated before. Set of feasible tasks which may be added to the workstation 1, T = [B, C].

From the set T, any one task will be selected based on the order of largest number of following tasks. From Table 2.1, C is to be given priority over B. Hence, C is selected for an addition to the workstation 1.

Remaining unassigned time = 40 – 35 = 5 seconds.

From Fig. 2.6, once C is assigned, the tasks which follow it are D and E. At the same time, task B may also be considered because there is no such constraint as task B can not be performed after C.

Set of tasks which may be assigned (without checking the feasibility),

S = [B, D, E].

Unassigned time is only 5 seconds whereas B, D and E each takes more than that.

Set of feasible tasks, T = [Nil].

Workstation 1 can not accommodate any more task.

This includes tasks A and C.

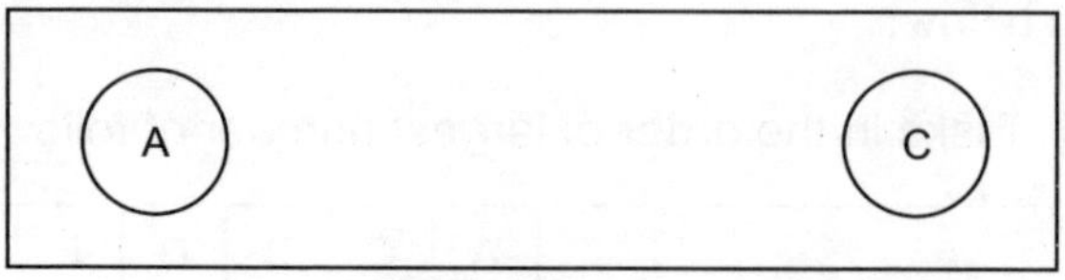

Workstation 1

Workstation 2

Set of tasks, S, 'which was recently obtained' = [B, D, E]. From this set, the first task for workstation 2 is B because it is prior to D & E as can be observed from Table 2.1.

Workstation 2 can also accommodate tasks whose total time does not exceed the cycle time, *i.e.* 80 seconds.

Time which is yet to be assigned = 80 – 30 = 50 seconds.

Now S = [D, E, F].

Set of feasible tasks, T = [D, E, F].

From Table 2.1, D, E and F have similar number of following tasks, *i.e.* 2. There is a tie in applying the largest number of following tasks rule. This is resolved by selecting the task with the maximum operation time. Task time for D, E and F are 30, 20, and 45 seconds respectively. As the task F takes maximum time, it is to be selected.

[Note: In case of a tie further, any one task may be selected arbitrarily.]

Now remaining time = 50 – 45

= 5 seconds.

S = [D, E, H] and as each task takes more than 5 seconds, T = [Nil].

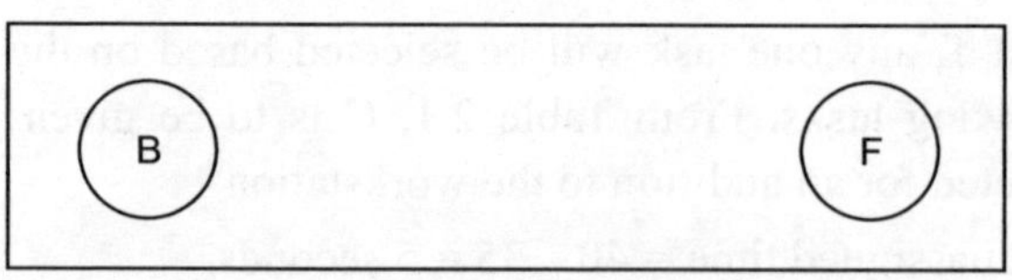

Workstation 2

Workstation 3

Recently obtained, S = [D, E, H]

First task to be selected for workstation 3 is based on the largest number of following tasks. But there is a tie between D and E. Task time for D, *i.e.* 30 seconds is more than that of E, therefore D is selected.

Remaining time = 80 – 30

= 50 seconds.

Again S = [E, H],

[Note: Although G is succeeding D, but it can not be considered unless E is over. As D and E both are predecessors to G, G can be taken into consideration only when both D and E are complete. (Refer Fig. 2.6)]

and T = [E, H].

E can easily be selected from Table 2.1 as discussed before.

Remaining time = 50 – 20

= 30 seconds.

Now S = [G, H],

and T = [H] as G takes 50 seconds which is more than the remaining time (30 seconds).

H is included in the workstation and as the remaining time = 30 – 30 = 0 second, no other task can be added.

However, the set of tasks without checking feasibility, S = [G].

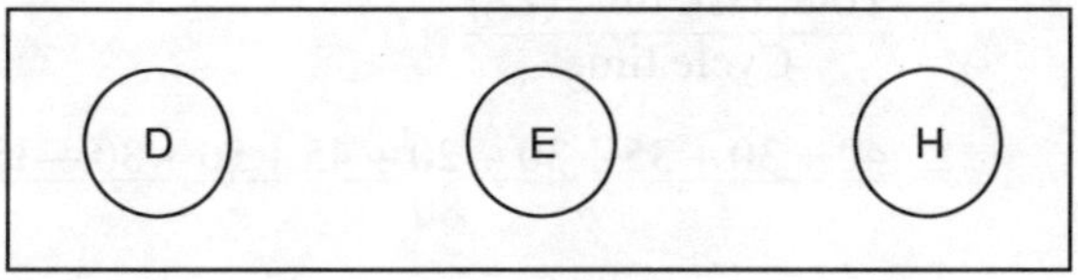

Workstation 3

Workstation 4

G is the first task for workstation 4.

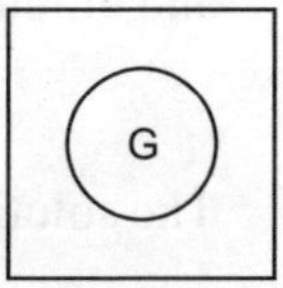

Workstation 4

Remaining time = 80 – 50

= 30 seconds.

S = [I], but T = [Nil].

Workstation 5

S = [I].

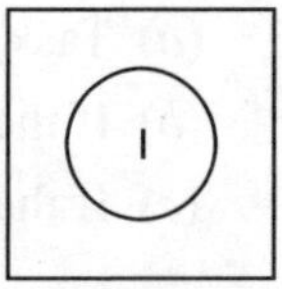

Workstation 5

Only one task, *i.e.* the last task I remains to be included in this workstation.

In this way, all the tasks have been accommodated in the five workstations. Figure 2.7 shows the workstations with one or more tasks. Time in seconds for all the tasks is also shown.

In order to evaluate the performance of the proposed solution, *i.e.* a creation of 5 workstations, the efficiency needs to be calculated.

$$\text{Efficiency (\%)} = \frac{\text{Theoretical minimum number of workstations} \times 100}{\text{Actual number of workstations obtained}}$$

Here, actual number of workstations obtained = 5,

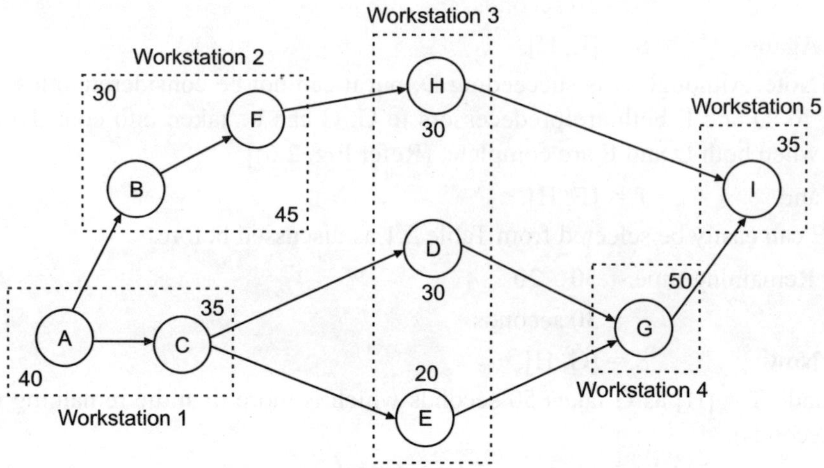

Fig. 2.7 Workstations including one or more tasks.

and the theoretical minimum number of workstations

$$= \frac{\text{Total task time } (\Sigma t_1)}{\text{Cycle time}}$$

$$= \frac{40 + 30 + 35 + 30 + 20 + 45 + 50 + 30 + 35}{80}$$

$$= \frac{315}{80}$$

$$= 3.9375$$

and efficiency of the proposal $= \dfrac{3.9375}{5} \times 100$

$$= 78.75\%$$

The solution process is briefly explained in Table 2.2.

Summary of the Largest Number of following Tasks Rule

1. Selection of the task, whether it is the first task in the workstation or subsequent tasks, will be made as follows:
   (*a*) Task with largest number of following tasks is selected.
   (*b*) If there is a tie in (*a*), task with maximum operation time is selected.
   (*c*) If there is a tie in (*b*), any task is selected arbitrarily.

Table 2.2 Assembly line balancing using the largest number of following tasks rule

	Tasks	Time	Time yet to be assigned	Set of tasks, S	Set of feasible tasks, T	Task with largest number of followers	Task with maximum operation time	Selected first task for the next workstation
Workstation 1	A	40	40	[B,C]	[B,C]	C	–	–
	C	35	5	[B, D, E]	[Nil]	–	–	B
Workstation 2	B	30	50	[D,E,F]	[D,E,F]	D,E,F,	F	–
	F	45	5	[D,E,H]	[Nil]	–	–	D
Workstation 3	D	30	50	[E,H]	[E,H]	E	–	–
	E	20	30	[G,H]	[H]	H	–	–
	H	30	0	[G]	[Nil]	–	–	G
Workstation 4	G	50	30	[I]	[Nil]	–	–	I
Workstation 5	I	35	45	–	–	–	–	–

2. Workstation is a physical location where one or more tasks are performed. Workstation should be able to complete most assigned tasks in a cycle time where,

$$\text{Cycle time} = \frac{\text{Production time per period}}{\text{Required production quantity per period}}$$

3. Precedence diagram is constructed.
4. The rule is implemented to balance the assembly line in the following steps:
   (*a*) Select the first task as explained in (1) above, and assign it to the workstation 1. Compute the time yet to be assigned.
   (*b*) Find the set of tasks, S, considering the precedence diagram only, *i.e.* without checking the feasibility with respect to time. From S, obtain the set of feasible tasks, T, by comparing each task time with the unassigned time. Select a task from T as discussed in 1.
   (*c*) If the set of feasible tasks, T, becomes empty, then no more tasks can be accommodated in the current workstation. Select one task from the recently obtained set, S, as described in (1). Selected task becomes the first assigned task for the next workstation. Compute the time yet to be assigned.
   (*d*) Repeat (*b*) and (*c*) until all the tasks are assigned.
5. To evaluate the performance of the proposed solution, Efficiency (%)

$$= \frac{\text{Theoretical minimum number of workstations}}{\text{Actual number of workstations obtained (4)}} \times 100$$

Whereas theoretical number of workstations,

$$= \frac{\text{Total task time } (\Sigma ti)}{\text{Cycle time}}$$

and ti = time taken by task i

Another heuristic to balance the assembly line is the longest operation time heuristic. In this rule, a task is selected primarily on the basis of the longest operation time first. In case of a tie, the task with the largest number of followers is selected.

2.5.2 Incremental Utilization Heuristic

In this procedure, the worker utilization is considered. The tasks are simply added one at a time to the workstation until the utilization of workers is 100% or it begins to decrease.

It is assumed that each worker in the group employed at a workstation does all the tasks assigned to that workstation. For example, the total task time for a workstation is 0.3 minutes (let us say, two tasks A and B are assigned to the workstation with the task time 0.16 and 0.14 min. respectively. Thus total task time is 0.16 + 0.14 = 0.3 min. If the cycle time is 0.1 min., then the number of workers required at the workstation would be 0.3 ÷ 0.1 = 3 and the worker utilization would be 100%.

Theoretical minimum number of workers for a workstation

$$= \frac{\text{Total task time for workstation}}{\text{Cycle time}}$$

If a third task, C, with the time 0.14 min. is added to the workstation, then the total task time for it will be 0.16 + 0.14 + 0.14 = 0.44 min. The theoretical minimum number of workers required for the workstation would be 0.44 ÷ 0.1 = 4.4. But the actual number of workers will be an integer, *i.e.* 5.

Worker Utilization (%)

$$= \frac{\text{Theoretical minimum number of workers}}{\text{Actual number of workers}} \times 100$$

$$= \frac{4.4}{5} \times 100$$

$$= 88\%.$$

Tasks are added to the workstation as per the precedence diagram. With the addition of each task, the theoretical and actual number of workers and subsequently worker utilization is computed. The tasks are added to the workstation until utilization of workers is 100% or the worker utilization decreases. If the utilization of workers is decreased by the addition of a task, then the particular task is excluded from the workstation and it becomes the first task to be included in the next workstation.

Example 2.2

An assembly line operates for 420 minutes per day. 2100 products are required to be assembled each day. The time taken by each task and the tasks which immediately precede each task are given below:

Tasks	Tasks which Immediately Precede	Time Taken (minutes)
A	-	0.63
B	A	0.17
C	B	0.82
D	C	0.91
E	C	0.16
F	D,E	0.33
G	F	0.66
H	G	0.21

(*a*) Draw the precedence diagram.

(*b*) Calculate the cycle time.

(*c*) Calculate the theoretical minimum number of workers for the whole assembly line.

(*d*) In order to minimize the workers' idle time, assign tasks to the workstations using the incremental utilization heuristic.

(*e*) Evaluate the proposed solution by computing efficiency of the proposal.

Now,

(*a*) The precedence diagram is as follows:

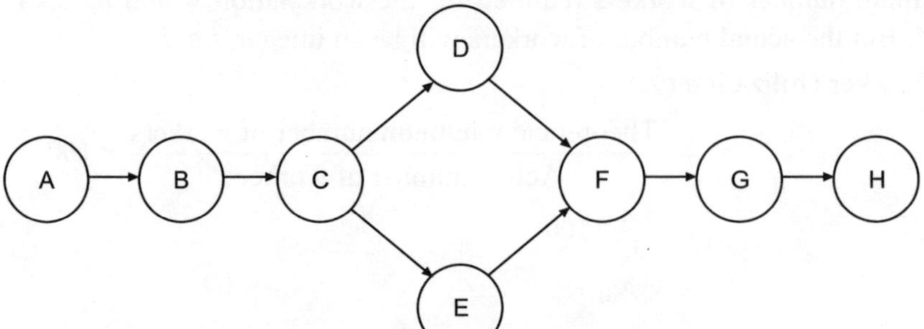

(*b*) $$\text{Cycle time} = \frac{\text{Production time per day}}{\text{Required production quantity per day}}$$

$$= \frac{420}{2100}$$

$$= 0.2 \text{ min.}$$

(*c*) Theoretical minimum number of workers for the whole assembly line

$$= \frac{\text{Total task time for the complete assembly}}{\text{Cycle time}}$$

$$= \frac{0.63 + 0.17 + 0.82 + 0.91 + 0.16 + 0.33 + 0.66 + 0.21}{0.2}$$

$$= \frac{3.89}{0.2}$$

$$= 19.45.$$

(*d*) Table 2.3 shows the computation for an assembly line balancing using the incremental utilization heuristic.

First task in the precedence diagram is A which is included in the workstation 1.

Task time of A = 0.63 min.

If the workstation 1 comprises only one task, A, then the theoretical number of workers required for this workstation

$$= \frac{0.63}{\text{Cycle time}}$$

$$= \frac{0.63}{0.2}$$

$$= 3.15.$$

As the theoretical number of workers is 3.15, the actual number of workers for workstation 1 (if it has only task A) = 4 and the worker utilization is

$$= \frac{3.15}{4}$$

$$= 0.7875 \text{ or } 78.75\%.$$

Table 2.3 Assembly line balancing by the incremental utilization heuristic

Tasks	A	A+B	C	C+D	C+D+E	E	E+F	E+F+G	E+F+G+H
Total task time (a)	0.63	.63+.17=0.8	.82	.82 + .91 = 1.73	1.73 + .16 = 1.89	.16	.16 + .33 = 0.49	.49 + .66 = 1.15	1.15 + .21 = 1.36
Theoretical No. of workers, b = (a)/ Cycle time	3.15	4	4.1	8.65	9.45	0.8	2.45	5.75	6.8
Actual No. of workers, c	4	4	5	9	10	1	3	6	7
Workers Utilization (%) = (b/c) × 100	78.75%	100%	82%	96.11%	94.5%	80%	81.67%	95.83%	97.14%
	Workstation 1		Workstation 2			Workstation 3			

Next task to be considered in the workstation is B. Total task time now for A + B is .63 + .17 = 0.8 min. Theoretical and actual number of workers is 4 each and the workers utilization is 100%. As the utilization is 100%, there is no need to add tasks further and the workstation 1 includes the tasks A and B with the requirement of 4 workers.

(A) + (B)	Theoretical number of workers = 4
Workstation 1	Actual number of workers = 4

To begin for workstation 2, the next task is C with time .82 minutes.

The theoretical number of workers = $\frac{0.82}{0.2} = 4.1$.

Actual number of workers would be the next higher integer, *i.e.* 5 and the worker utilization is

$$= \frac{4.1}{5} \times 100 = 82\%.$$

Next task to be added is either D or E as per the precedence diagram. Let us say D (in alphabetical order). Following the procedure as discussed before, the theoretical and actual numbers of workers are 8.65 and 9 respectively. The worker utilization is 96.11%. By adding the task E, utilization is 94.5%. As the worker utilization is decreased by an addition of task E, this task is excluded from the current workstation. Now workstation 2 has the tasks C & D.

(C) + (D)	Theoretical number of workers = 8.65
Workstation 2	Actual number of workers = 9

Task E was excluded for the workstation 2 and it becomes the first task for workstation 3, and similar procedure is repeated. Now tasks are added from F to H one-by-one and the worker utilization is observed to be increasing continuously to 97.14%. As no other task remains to be considered, the workstation 3 will consist of four tasks, E, F, G and H with the worker utilization 97.14%.

(E) + (F) + (G) + (H)	Theoretical number of workers = 6.8
Workstation 3	Actual number of workers = 7

Theoretical number of workers for the entire assembly line

= theoretical number of workers for the workstations 1, 2 and 3

= 4 + 8.65 + 6.8

= 19.45, which is similar to the number obtained in part (c) of the current example.

Actual number of workers for the entire assembly line

= actual number of workers for the workstations 1, 2 and 3

= 4 + 9 + 7

= 20

(*e*) Efficiency of the proposed solution

$$= \frac{\text{Theoretical total number of workers}}{\text{Actual total number of workers}}$$

$$= \frac{19.45}{20}$$

$$= 0.9725$$

$$= 97.25\%.$$

2.6 Planning for Process Layout

The main objective in designing the process or functional layout is to minimize the total distance travelled by the material or to minimize the material handling cost. Figure 2.8 provides a brief classification of the methods and computer programmes available for analyzing and developing the process type of layout.

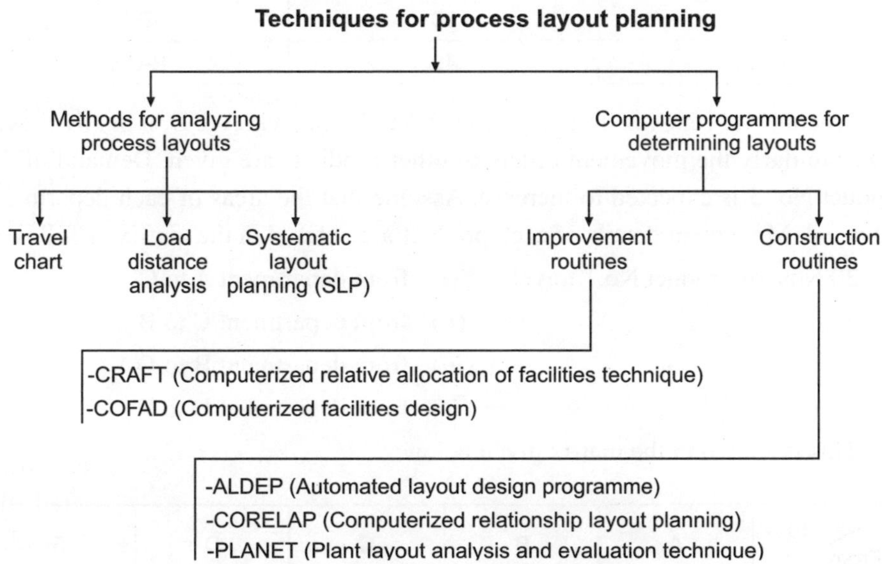

Fig. 2.8 Classification of techniques for planning the process layouts

2.6.1 Travel Chart

The travel chart is used to analyze the movement of material/components from department to department. Department refers to a place or location where either similar machines are installed or a kind of service is provided. The travel chart method will provide a good solution in the form of the arrangement of departments which is subject to further analysis and modification. This will take into account an assessment of flow or movement of parts or components in a specified period, and the departments, between which the movement is more, are to be positioned together or closer to each other as far as possible.

Example 2.3

A job shop is in the process of planning for the layout consisting of 5 departments, A, B, C, D and E. An idea is to place the departments in the order as shown in Fig. 2.9 where O is the space reserved for administrative office.

A	B	C
O	E	D

Fig. 2.9 An initial idea of the layout

Estimated production quantities per week and the movement order of 5 products are shown as follows:

Product No.	Production Quantity	Movement Order
1	20	A-C-B-D-E
2	30	D-A-B-C
3	25	A-D
4	60	C-D-E-A
5	40	B-C-D

Product No. 1 moves from department A to C, then via B & D, it finally moves to E. Similarly the movement orders of other products are given. Demand of the product No. 3 is expected to increase. Assume that the areas of each department are equal. Movement details of each product are entered in the matrix as follows:

20 Nos. of product No.1 travel (*i*) from department A to C

(*ii*) from department C to B

(*iii*) from department B to D

(*iv*) from department D to E

This is shown in the matrix given below:

To→ ↓From	A	B	C	D	E
A	–		20		
B		–		20	
C		20	–		
D				–	20
E					–

In this manner, the movement details of the remaining 4 products are entered in the matrix and finally the travel chart is as shown below in Fig. 2.10.

To→ / ↓From	A	B	C	D	E
A	–	30	20	25	
B		–	30 + 40 = 70	20	
C		20	–	60 + 40 = 100	
D	30			–	20 + 60 = 80
E	60				–

Fig. 2.10 Travel chart

From Fig. 2.10, the movements between two departments are as follows:

A–B	30
A–C	20
A–D	25*
B–C	70**
B–D	20
C–B	20**
C–D	100
D–A	30*
D–E	80
E–A	60

As we are interested in the total flow of material between departments, it is immaterial whether the products move from A to D or from D to A. Therefore 25* and 30* may be added to get the total flow between D–A/A–D. Similarly between B–C/C–B, the total movement is 70** + 20** = 90. Accordingly, the total movement between departments:

A–B	30
A–C	20
A–D	55
B–C	90
B–D	20
C–D	100
D–E	80
E–A	60

Arranging in the decreasing order of the total movement:

C–D	100
B–C	90
D–E	80

E–A	60
A–D	55
A–B	30
A–C	20
B–D	20

Refer to Fig. 2.9, C–D, B–C and D–E are closer to each other. As the above list is in the order of priority for the closeness of departments, an initial idea of the layout is suitable with reference to C–D, B–C and D–E. E–A are also comparatively closes to each other. But A–D are far from each other whereas A–B are closer. A–D must be given the priority over A–B as the total flow between A–D is 55 which is much higher than 30 between A–B. Further demand of the product 3 is expected to increase and the path followed by this product is A–D. It is reasonable to interchange the location of B and D which is shown in Fig. 2.11.

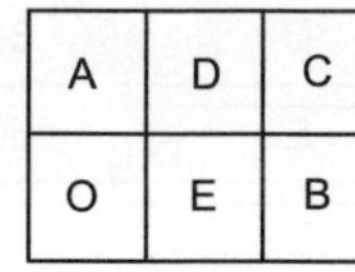

Fig. 2.11 Proposed layout

In this method, the distance between departments is not taken into consideration. Either it may be included in the travel chart analysis or it may be considered in the load-distance method which is explained in the next section.

2.6.2 Load-Distance Analysis

In many situations, the objective is to compare the two or more alternative layouts. If some alternatives are available, then depending on the movement of various products, the total distance travelled by all the items to be manufactured is computed for the alternative layouts. The preferred layout is obviously one in which the total distance travelled by the items is the least.

Example 2.4

After careful considerations, an organization has found the two alternative layouts X and Y as shown below:

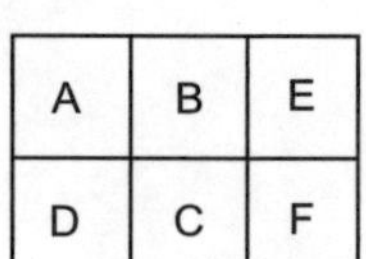

Layout - X

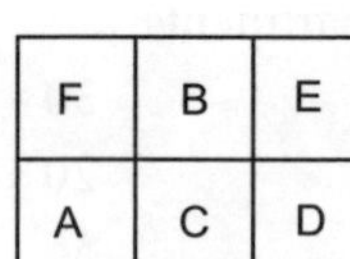

Layout - Y

Estimated production quantity per week and the movement order of three items to be produced are given below:

Item No	Production Quantity	Movement Order
1	200	A-B-D-F-E
2	150	D-E-F-C-A
3	100	F-C-B-A-E

Distances in (m) between departments have been provided for both the layouts as follows:

Department (From-To)	Distance (m)	
	Layout X	Layout Y
A-B/B-A	10	20
B-D	20	20
D-F	20	30
F-E/E-F	10	20
D-E	30	10
F-C	10	20
C-A	20	10
C-B	10	10
A-E	20	30

Now the distances moved by each product for both the layout alternatives are computed as follows:

Item	Movement Order	Distance Moved by Each Item (m)	
		Layout - X	Layout - Y
1	A-B-D-F-E	10 + 20 + 20 + 10 = 60	20 + 20 + 30 + 20= 90
2	D-E-F-C-A	30 + 10 +10 +20 = 70	10 + 20 + 20 + 10 = 60
3	F-C-B-A-E	10 + 10 + 10 + 20 = 50	20 + 10 + 20 + 30 = 80

Distance travelled by all the quantities of each item per week can be obtained as:

Item	Quantity	Distance Moved (m)	
		Layout - X	Layout - Y
1	200	60 x 200 = 12000 m	90 x 200 = 18000 m
2	150	70 x 150 = 10500 m	60 x 150 = 9000 m
3	100	50 x 100 = 5000 m	80 x 100 = 8000 m

Total movement (m) for each alternative layout:

Layout X: 12000 + 10500 + 5000 = 27500 m

Layout Y: 18000 + 9000 + 8000 = 35000 m

As the total movement for layout X is less, it is the preferred layout between alternatives.

2.6.3 Systematic Layout Planning (SLP)

In certain situations, when the flow of materials between departments is not the only significant factor on which a layout decision can be based, the systematic layout planning may be useful. Depending on several factors such as common equipment, common worker, normal flow sequence, better supervision, etc., closeness ratings are developed between departments. Considering the closeness rating, the two departments may need to be placed together because the pair of departments has absolutely necessary rating. But another pair of departments may have an undesirable rating and these should be located far from each other. Intermediate ratings are also developed based on various factors. Initial layout is considered and through trial and error, it is modified by placing departments on the basis of their closeness ratings subject to space constraints.

2.6.4 CRAFT (Computerized Relative Allocation Facilities Technique)

As shown in Fig. 2.8, this is an improvement routine. Improvement routines are those algorithms which require an initial layout of facilities. Objective of the CRAFT is to minimize the total material handling cost. In this heuristic procedure, an improvement in the layout is made by interchanging the departments and comparing the costs. The method is discussed briefly as follows:

1. Input details are (*a*) Initial layout, (*b*) Cost of material handling per unit distance and (*c*) Movement data. Compute the total handling cost corresponding to an initial layout.
2. Compute the cost of handling with each pair-wise interchange of the centroids of departments and select the least cost.
3. If the least cost obtained in (2) is less than the cost of the previous layout, make the actual interchange of the selected pair of departments and compute the cost of the revised layout.
4. Repeat the procedure from (2) onwards as long as an improvement in the cost is observed.

Following are the assumptions in the CRAFT:

(*a*) Distance between two departments is the rectilinear distance between the centroids of the departments. For example, two departments A and B are adjacent to each other.

The bottommost horizontal line may be considered as X-axis and the leftmost vertical line is treated as Y-axis in the diagram.

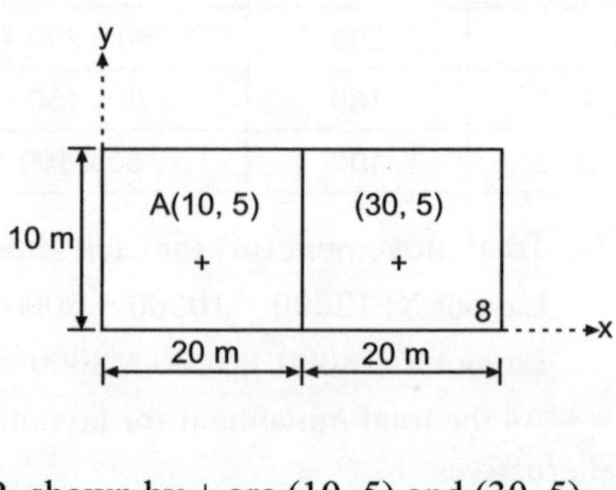

Coordinates of the centroids of A and B, shown by + are (10, 5) and (30, 5).

The distance between centroids is 20 m which is obtained as:

$$|10 - 30| + |5 - 5| = |-20| + |0| = 20 + 0 = 20 \text{ m}$$

(ignoring the negative sign of |–20|).

If centroids of the departments A and B are denoted as (x_A, y_A), and (x_B, y_B), then the distance between A and B is considered as:

$$d_{AB} = |x_A - x_B| + |y_A - y_B|$$

In case the departments A and B are as shown below, then the centroids are $(x_A, y_A) = (5, 5)$ and $(x_B, y_B) = (20, 10)$. The distance between two departments is assumed to be the rectilinear distance (shown by dotted lines) between centroids.

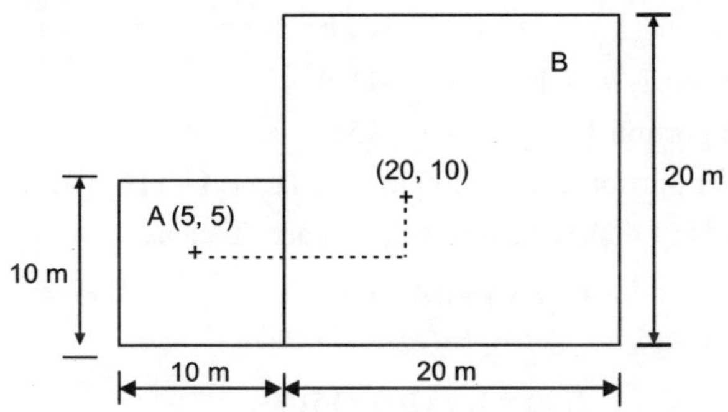

$$\begin{aligned} d_{AB} &= |5 - 20| + |5 - 10| \\ &= |-15| + |-5| \\ &= 15 + 5 = 20 \text{ m.} \end{aligned}$$

If a department does not have the rectangular or square shape, then the centriod of the department is obtained as follows:

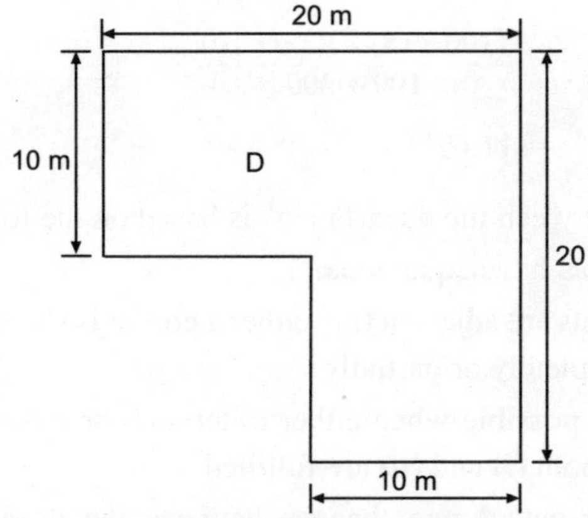

A department D is shown in the above fig. The objective is to find out its centroid. The whole area should be divided in such a way that the regular shapes are obtained as shown below:

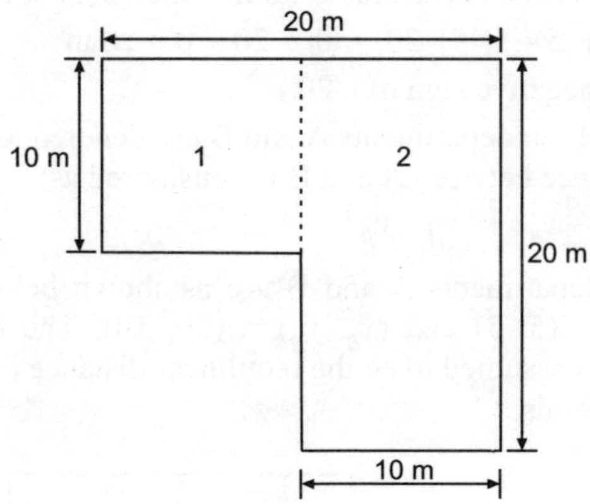

Area of portion 1, $a_1 = 10 \times 10 = 100\ m^2$.

Centroid of portion 1, $(x_1, y_1) = (5, 15)$.

Similarly, for portion 2, $a_2 = 200\ m^2$ and $(x_2, y_2) = (15, 10)$.

The centroids of department D, (x_D, y_D) are given as

$$x_D = \frac{(a_1 \times x_1) + (a_2 \times x_2)}{a_1 + a_2}$$

$$= \frac{(100 \times 5) + (200 \times 15)}{100 + 200}$$

$$= \frac{500 + 3000}{300}$$

$$= 11.67$$

and

$$y_D = \frac{(a_1 \times y_1) + (a_2 \times y_2)}{a_1 + a_2}$$

$$= \frac{(100 \times 15) + (200 \times 10)}{100 + 200}$$

$$= 11.67$$

(*b*) Interchange between the departments is based on the following criteria:

(*i*) Departments have equal area.

(*ii*) Departments are adjacent (*i.e.* either a border is shared by the departments either completely or partially).

Interchange is possible when either criteria (*i*) or criteria (*ii*) is satisfied, as well as when both (*i*) and (*ii*) are fulfilled.

(*c*) While making actual interchanges between the departments, if it is not possible to retain the shape of any of the department, then shape may change. However obviously, area of the department should remain the same.

Example 2.5

In order to implement the CRAFT, an initial layout is given as follows:

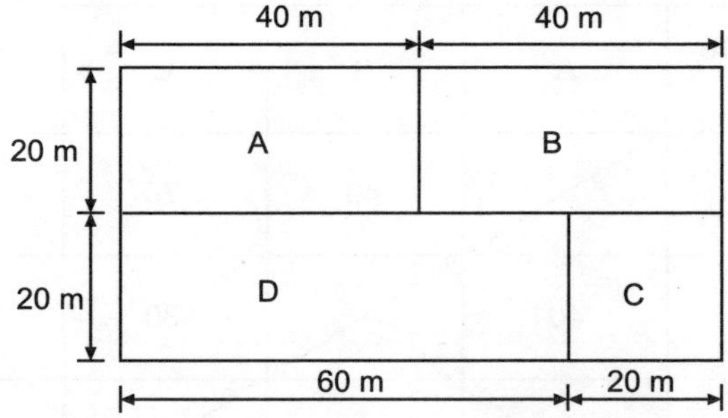

Fig. 2.12 Initial Layout to implement CRAFT

Four departments, A, B, C & D are to be located in the total available area, $80 \times 40 = 3200$ m^2. Load L_{ij} is to be transported from one department i to another j in a week is given in the movement matrix shown below:

From↓ \ To→	A	B	C	D
A		1	3	2
B	4		2	3
C	2	3		4
D	0	1	2	

Fig. 2.13 Movement matrix

Suppose that, instead of transportation of one item at a time, the number of items are moving between departments on a single trip, then the movement matrix may also refer to the number of trips for material handling per week or any other specified time period.

As discussed before, the centroids of departments may be obtained as given below:

x_A, y_A	x_B, y_B	x_C, y_C	x_D, y_D
20, 30	60, 30	70, 10	30, 10

Distance between the departments, *A* and *B*,

$$d_{AB} = |x_A - x_B| + |y_A - y_B|$$
$$= |20 - 60| + |30 - 30| = 40 \text{ m}$$

Similarly the distances between each pair of departments are computed, and these are shown in Fig. 2.14.

From \ To	A	B	C	D
A		40	70	30
B	40		30	50
C	70	30		40
D	30	50	40	

Fig. 2.14 Distance in m

As the distance between A–B, A–C, A–D is similar to B–A, C–A, D–A respectively, the first row on the right side of the diagonal is similar to the first column on the left side of the diagonal in Fig. 2.14. Similarly the second and third columns on the left side will be the same as second and third rows on the right side of diagonal respectively.

If there are n departments ($n = 4$ in this example),

$$\text{Total movement } (m) = \int_{i=1}\int_{j=1} L_{ij}\, d_{ij}$$

$$\text{Total cost of transportation} = \int_{i=1}\int_{j=1} Li_j\, d_{ij}\, K_{ij}$$

Where K_{ij} is the cost per m of transportation from the department i to j.

If the cost per unit m is constant, say K irrespective of the pair of department,

$$\text{the total material handing/transportation cost} = K \int_{i=1}\int_{j=1} L_{ij}\, d_{ij}$$

As K is constant, minimizing the total cost would be similar to minimizing the total movement/distance travelled. For simplicity, K is assumed as ₹ 1 per m so that the numerical value of total cost will also be similar to the total distance moved.

For ease in computation, Fig. 2.13 is superimposed on Fig. 2.14 as shown in Fig. 2.15. Load to be moved or the number of trips per period, L_{ij} are shown in the left comer of each cell.

To→ From↓	A	B	C	D
A		1 40	3 70	2 30
B	4 40		2 30	3 50
C	2 70	3 30		4 40
D	0 30	1 50	2 40	

Fig. 2.15 Distance and movement between the departments

[Note: If K_{ij} is different for the different pair of departments, then it can be written in the right corner of each cell. Total cost of transportation will be the sum of multiplication of three numbers in each cell. In the present example, as K is constant, *i.e.* K = ₹ 1 per metre distance, it is ignored.]

$$\begin{aligned} \text{Total cost of the initial layout} &= 1 \times [(40 \times 1) + (70 \times 3) + (30 \times 2) + (40 \times 4) \\ &\quad + (30 \times 2) + (50 \times 3) + (70 \times 2) + (30 \times 3) \\ &\quad + (40 \times 4) + (30 \times 0) + (50 \times 1) + (40 \times 2)] \\ &= ₹\ 1200.00. \end{aligned}$$

Now the procedure is to interchange the centroids of departments and compare the least cost corresponding to an interchange with ₹ 1200. If it is less than ₹ 1200 then make the actual interchange and compute the cost.

Cycle 1

Interchange between departments may be as follows:

A–B
A–C
A–D
B–C
B–D
C–D

As the interchange between A and C does not satisfy any of the criteria, *i.e.* A and C do not have equal area and they are also not adjacent, therefore A–C will be dropped. A and B are adjacent as well as both have equal area, *i.e.* 800 m^2 and A–B will be considered for the interchange. Remaining departments will also be interchanged because these are adjacent. While interchanging the departments, initially only centroids will be interchanged. An actual interchange will be made only for the pair which has the potential of reduction in cost.

For interchange of the centroids, the reference set of centroids for the existing (initial) layout is as follows:

x_A, y_A	x_B, y_B	x_C, y_C	x_D, y_D
20, 30	60, 30	70,10	30,10

Fig. 2.16 Reference set of centroids for cycle 1

Interchange A–B

As discussed, the centroid of A and B will take the place of each other. Now the position of centroids for this interchange is as follows:

x_A, y_A	x_B, y_B	x_C, y_C	x_D, y_D
60, 30	20, 30	70, 10	30, 10

Distance between each pair of the departments is obtained and the movement data are superimposed as discussed before:

From \ To	A	B	C	D
A		1 40	3 30	2 50
B	4 40		2 70	3 30
C	2 30	3 70		4 40
D	0 50	1 30	2 40	

The corresponding cost

$$= (40 \times 1) + (30 \times 3) + (50 \times 2) + (40 \times 4) + (70 \times 2) + (30 \times 3) + (30 \times 2) + (70 \times 3) + (40 \times 4) + (50 \times 0) + (30 \times 1) + (40 \times 2)$$

$$= ₹\ 1160.00.$$

Interchange A-D

From the reference set of centroids shown in Fig. 2.16, interchange the centroid of A and D as follows:

x_A, y_A	x_B, y_B	x_C, y_C	x_D, y_D
30, 10	60, 30	70, 10	20, 30

From \ To	A	B	C	D
A		1 50	3 40	2 30
B	4 50		2 30	3 40
C	2 40	3 30		4 70
D	0 30	1 40	2 70	

and the corresponding cost = ₹ 1240.

Interchange B–C

Again refer to the Fig. 2.16 and the interchange is as follows:

x_A, y_A	x_B, y_B	x_C, y_C	x_D, y_D
20, 30	70, 10	60, 30	30, 10

From \ To	A	B	C	D
A		1 70	3 40	2 30
B	4 70		2 30	3 40
C	2 40	3 30		4 50
D	0 30	1 40	2 50	

Cost = ₹ 1220.

Interchange B–D

x_A, y_A	x_B, y_B	x_C, y_C	x_D, y_D
20, 30	30, 10	70, 10	60, 30

From \ To	A	B	C	D
A	—	1 / 30	3 / 70	2 / 40
B	4 / 30	—	2 / 40	3 / 50
C	2 / 70	3 / 40	—	4 / 30
D	0 / 40	1 / 50	2 / 30	—

Cost = ₹ 1160.

Interchange C–D

x_A, y_A	x_B, y_B	x_C, y_C	x_D, y_D
20, 30	60, 30	30, 10	70, 10

From \ To	A	B	C	D
A	—	1 / 40	3 / 30	2 / 70
B	4 / 40	—	2 / 50	3 / 30
C	2 / 30	3 / 50	—	4 / 40
D	0 / 70	1 / 30	2 / 40	—

Cost = ₹ 1100.

Summary of the cost corresponding to interchange is given below:

Interchange:	**(A–B)**	**(A–D)**	**(B–C)**	**(B–D)**	**(C–D)**
Cost (₹):	1160	1240	1220	1160	1100

Minimum cost is ₹ 1100 corresponding to the change of centroids of departments C–D. As this is less than the cost of the existing layout, *i.e.* ₹ 1200, make the actual interchange of department C–D.

Refer to Fig. 2.12 and interchange the location of C and D as shown in Fig. 2.17.

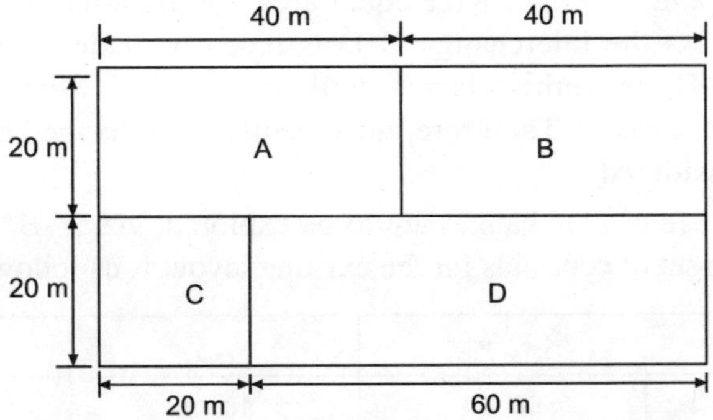

Fig. 2.17 Revised Layout at the end of cycle 1

While making actual interchange, centroids of the departments may change, therefore recalculate the centroids considering Fig. 2.17 and then obtain the actual total cost.

x_A, y_A	x_B, y_B	x_C, y_C	x_D, y_D
20, 30	60, 30	10, 10	50, 10

From \ To	A	B	C	D
A		1 40	3 30	2 50
B	4 40		2 70	3 30
C	2 30	3 70		4 40
D	0 50	1 30	2 40	

Total actual cost of the revised layout = ₹ 1160.

If this cost is less than the cost of the existing layout, then only retain Fig. 2.17 as the revised layout. As in this case, the total actual cost ₹ 1160 is less than ₹ 1200, the revised layout which will be considered for further computation is retained and it will replace the initial layout. Now call it as the existing layout. For comparing the cost and interchange between the departments, the recently obtained layout, centroids and cost will act as the reference. Now the next cycle is to be followed.

Cycle 2

Out of the interchanges A–B, A–C, A–D, B–C, B–D and C–D, since the departments B and C neither have equal area nor are adjacent, B–C is not to be considered. As the interchange C–D is recently made, if we try to again interchange–C–D, the initial layout will be obtained which will yield no improvement in the cost. Therefore, the recently interchanged department pair is not to be considered.

For cycle 2, four interchanges are to be explored, viz. A–B, A–C, A–D and B–D. Reference set of centroids for the existing layout is as follows:

x_A, y_A	x_B, y_B	x_C, y_C	x_D, y_D
20, 30	60, 30	10, 10	50, 10

Interchange A – B

x_A, y_A	x_B, y_B	x_C, y_C	x_D, y_D
60, 30	20, 30	10, 10	50, 10

From \ To	A	B	C	D
A		1 40	3 70	2 30
B	4 40		2 30	3 50
C	2 70	3 30		4 40
D	0 30	1 50	2 40	

Cost = ₹ 1200.

Interchange A–C

x_A, y_A	x_B, y_B	x_C, y_C	x_D, y_D
10, 10	60, 30	20, 30	50, 10

To / From	A	B	C	D
A		1 70	3 30	2 40
B	4 70		2 40	3 30
C	2 30	3 40		4 50
D	0 40	1 30	2 50	

Cost = ₹ 1200.

Interchange A–D

x_A, y_A	x_B, y_B	x_C, y_C	x_D, y_D
50, 10	60, 30	10, 10	20, 30

To / From	A	B	C	D
A		1 30	3 40	2 50
B	4 30		2 70	3 40
C	2 40	3 70		4 30
D	0 50	1 40	2 30	

Cost = ₹ 1140.

Interchange B–D

x_A, y_A	x_B, y_B	x_C, y_C	x_D, y_D
20, 30	50, 10	10, 10	60, 30

From \ To	A	B	C	D
A		1 / 50	3 / 30	2 / 40
B	4 / 50		2 / 40	3 / 30
C	2 / 30	3 / 40		4 / 70
D	0 / 40	1 / 30	2 / 70	

Cost = ₹ 1220.

Out of the above interchanges, the minimum cost, is ₹ 1140 corresponding to A–D which is less than the cost of the existing layout (₹ 1160). Actual interchange between A and D in the existing layout (Fig. 2.17) will yield the following revised layout at the end of cycle 2 as shown in Fig. 2.18.

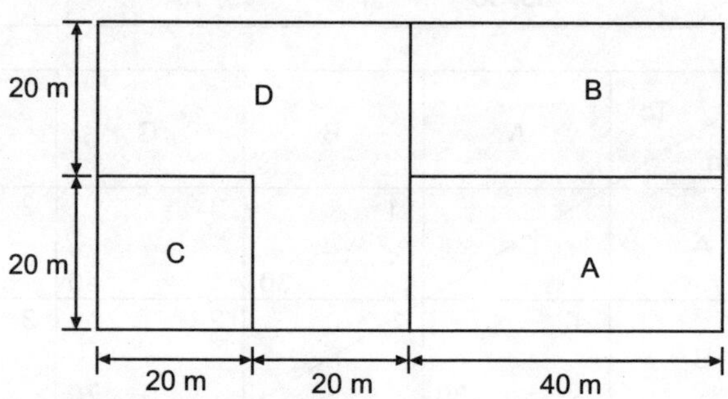

Fig. 2.18 Revised layout at the end of cycle 2

Refer to Fig. 2.17, A has taken the place of D as shown in Fig. 2.18. In the remaining unoccupied portion, D is accommodated. Area of D is the same as before, but the shape has changed as shown below:

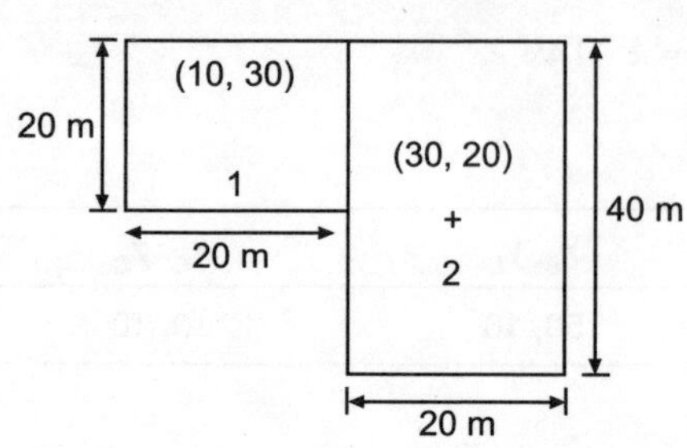

The whole area of department may be divided into two regular shapes, 1 and 2, in order to obtain the centroid (X_D, Y_D) as explained before. The centroids of portions, 1 and 2, are shown by + sign and the values are (10, 30) and (30, 20) respectively.

$$x_D = \frac{a_1 x_1 + a_2 x_2}{a_1 + a_2},$$

where $a_1 = 400\ \text{m}^2, a_2 = 800\text{m}^2, x_1 = 10, x_2 = 30,$

and $$x_D = \frac{(400 \times 10) + (800 \times 30)}{1200} = 23.33.$$

$$y_D = \frac{a_1 y_1 + a_2 y_2}{a_1 + a_2},$$

where $y_1 = 30,\ \ y_2 = 20,$

and $$y_D = \frac{(400 \times 30) + (800 \times 20)}{1200} = 23.33.$$

Now with reference to the Fig. 2.18, the centroids are as follows:

x_A, y_A	x_B, y_B	x_C, y_C	x_D, y_D
60, 10	60, 30	10, 10	23.33, 23.33

From \ To	A	B	C	D
A		1 / 20	3 / 50	2 / 50
B	4 / 20		2 / 70	3 / 43.44
C	2 / 50	3 / 70		4 / 26.66
D	0 / 50	1 / 43.34	2 / 26.66	

Total actual cost of the revised layout = ₹ 1133.32.

As it is less than the cost of the existing layout (₹ 1160), Fig. 2.18 is retained as the revised layout.

Cycle 3

As A–D is recently interchanged, it will not be considered.

A–C and B–C do not satisfy either of the criteria for interchange. Possibilities to be explored confine to the three pair-wise interchanges A–B, B–D and C–D.

Interchange A–B

Refer to the recently obtained centroids for the revised layout and interchange them for A–B.

x_A, y_A	x_B, y_B	x_C, y_C	x_D, y_D
60, 30	60, 10	10, 10	(23.33, 23.33)

From \ To	A	B	C	D
A		1 / 20	3 / 70	2 / 43.34
B	4 / 20		2 / 50	3 / 50
C	2 / 70	3 / 50		4 / 26.66
D	0 / 43.34	1 / 50	2 / 26.66	

Cost = ₹ 1146.64.

Interchange B–D

x_A, y_A	x_B, y_B	x_C, y_C	x_D, y_D
60, 10	23.33, 23.33	10, 10	60, 30

From \ To	A	B	C	D
A		1 / 50	3 / 50	2 / 20
B	4 / 50		2 / 26.66	3 / 43.34
C	2 / 50	3 / 26.66		4 / 70
D	0 / 20	1 / 43.34	2 / 70	

Cost = ₹ 1266.66.

Interchange C–D

x_A, y_A	x_B, y_B	x_C, y_C	x_D, y_D
60, 10	60, 30	(23.33, 23.33)	(10, 10)

From \ To	A	B	C	D
A		1 20	3 50	2 50
B	4 20		2 43.34	3 70
C	2 50	3 43.34		4 26.66
D	0 50	1 70	2 26.66	

Cost = ₹ 1106.66.

Minimum cost obtained is ₹ 1106.66 corresponding to the interchange C–D which is less than the cost of the existing layout (₹ 1133.32), therefore make the actual interchange C–D as shown in Fig. 2.19.

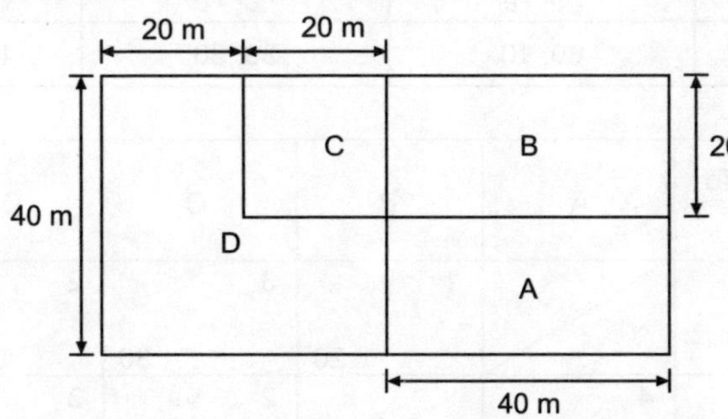

Fig. 2.19 Revised layout at the end of cycle 3

As discussed in cycle 2, the centriods are obtained as:

x_A, y_A	x_B, y_B	x_C, y_C	x_D, y_D
60, 10	60, 30	(30, 30)	16.66, 16.66

From \ To	A	B	C	D
A	—	1 / 20	3 / 50	2 / 50
B	4 / 20	—	2 / 30	3 / 56.68
C	2 / 50	3 / 30	—	4 / 26.68
D	0 / 50	1 / 56.68	2 / 26.68	—

Total actual cost of the revised layout = ₹ 986.80.

As it is less than the cost of the existing layout (₹ 1133.32), Fig. 2.19 is retained as the revised layout.

Cycle 4

Since C–D is recently interchanged and A–C and B–D do not satisfy either of the criteria, these pairs will not be considered.

Interchange A–B

x_A, y_A	x_B, y_B	x_C, y_C	x_D, y_D
60, 30	60, 10	30, 30	16.66, 16.66

From \ To	A	B	C	D
A	—	1 / 20	3 / 30	2 / 56.68
B	4 / 20	—	2 / 50	3 / 50
C	2 / 30	3 / 50	—	4 / 26.68
D	0 / 56.68	1 / 50	2 / 26.68	—

Cost = ₹ 973.44.

Interchange A–D

x_A, y_A	x_B, y_B	x_C, y_C	x_D, y_D
16.66, 16.66	60, 30	30, 30	60, 10

From \ To	A	B	C	D
A		1 56.68	3 26.68	2 50
B	4 56.68		2 30	3 20
C	2 26.68	3 30		4 50
D	0 50	1 20	2 50	

Cost = ₹ 1046.80.

Interchange B–C

x_A, y_A	x_B, y_B	x_C, y_C	x_D, y_D
60, 10	30, 30	60, 30	16.66, 16.66

From \ To	A	B	C	D
A		1 50	3 20	2 50
B	4 50		2 30	3 26.68
C	2 20	3 30		4 56.68
D	0 50	1 26.68	2 56.68	

Cost = ₹ 1046.80.

The least cost is ₹ 973.44 corresponding to A–B and it is also less than ₹ 986.80 (cost of the existing layout), therefore make actual interchange as shown in Fig. 2.20.

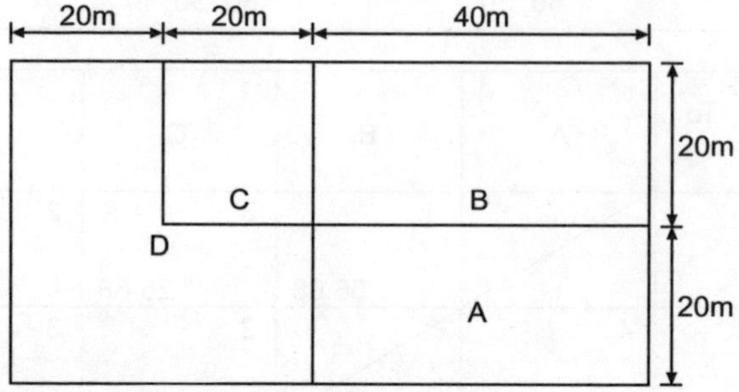

Fig. 2.20 Revised layout at the end of cycle 4

With the actual interchange of departments, the centroids are similar to the situation when only centroids of A and B were interchanged. Therefore the total actual cost of the revised layout = ₹ 973.44.

Cycle 5

As A–B is recently interchanged, and A–D and B–C need not be considered, Possibilities for further improvement in the cost are to be explored for interchange A–C, B–D and C–D as follows:

Interchange A–C

x_A, y_A	x_B, y_B	x_C, y_C	x_D, y_D
30, 30	60, 10	60, 30	16.66, 16.66

From \ To	A	B	C	D
A		1, 50	3, 30	2, 26.68
B	4, 50		2, 20	3, 50
C	2, 30	3, 20		4, 56.68
D	0, 26.68	1, 50	2, 56.68	

Cost = ₹ 1093.44.

Interchange B–D

x_A, y_A	x_B, y_B	x_C, y_C	x_D, y_D
60, 30	16.66, 16.66	30, 30	60, 10

Cost = ₹ 1106.80.

Interchange C–D

x_A, y_A	x_B, y_B	x_C, y_C	x_D, y_D
60, 30	60, 10	(16.66, 16.66)	30, 30

From \ To	A	B	C	D
A		1 / 20	3 / 56.68	2 / 30
B	4 / 20		2 / 50	3 / 50
C	2 / 56.68	3 / 50		4 / 26.68
D	0 / 30	1 / 50	2 / 26.68	

Cost = ₹ 1053.48.

Minimum cost is ₹ 1053.48 which is more than the cost of the existing layout (₹ 973.44). Hence the process of further computation is stopped and Fig. 2.20 represents the proposed layout with the total cost of ₹ 973.44.

2.6.5 COFAD (Computerized Facilities Design)

Computerized facilities design (COFAD) is further extension of CRAFT. Overall cost also depends on the selection of material handling equipments in addition to the distance travelled. COFAD considers various options for the material handling equipments from the point of view of equipment utilization and cost. After selecting the equipment, the layout is improved following the procedure of CRAFT.

2.6.6 ALDEP (Automated Layout Design Programme)

As shown in Fig. 2.8, ALDEP is a construction routine. Unlike improvement routines, an initial layout is not required. A relationship chart is developed which will show the relationship between departments in the form of closeness ratings. Closeness ratings are as follows:

Absolutely necessary (A): The letter code for 'absolutely necessary' closeness rating is 'A'. For example, if a similar facility or service centre is being used by the

		processes to be carried out in two departments, then these need to be adjacent to each other.
Especially important (E)	:	For example, if the same human resources are needed for two departments, then from the point of view of supervision, these may be considered under closeness rating 'E'.
Important (I)	:	It may be important to arrange the departments in the same sequence as the normal flow of material takes place.
Ordinary importance (O)	:	If it is not necessary to keep two departments together, but the management feels it convenient to place them near each other as far as possible.
Unimportant (U)	:	If we are indifferent about the location of two or more departments, then it is not a matter of consideration whether two departments should be close to each other or far from each other.
Undesirable (X)	:	From safety and other considerations, it may be undesirable to place the two departments adjacent to each other. This rating is denoted by 'X'.

Numerical values attached to the closeness ratings are as follows:

Letter Code	Numerical Value
A	$4^3 = 64$
E	$4^2 = 16$
I	$4^1 = 4$
O	$4^0 = 1$
U	0
X	$-4^5 = -1024$

First department is selected at random to begin with and then depending on the closeness rating, the next department is chosen to be adjacent to the first department. Suitable sweep width in number of columns needs to be specified.

For example, if a department, a, has an area of 10 sq.m, then a may be shown either as:

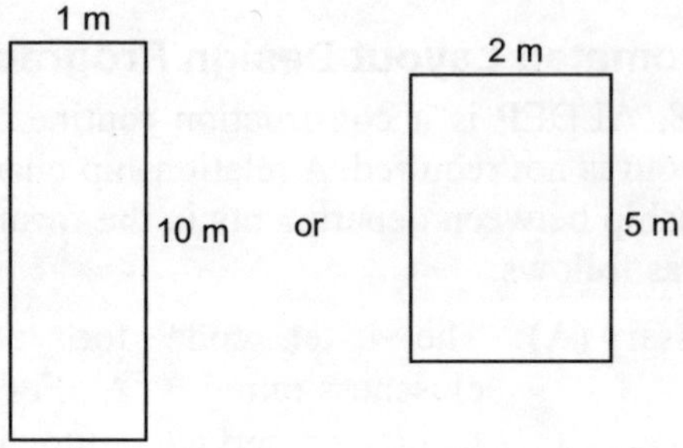

If the sweep width is 1 column, then starting from the top left comer of the layout, a is printed 10 times vertically as follows:

a
a
a
a
a Each 'a' represents 1 sq. m. area.
a
a
a
a
a

If the sweep width is 2 columns, then

aa
aa
aa
aa
aa

Usually odd shapes of the departments are obtained in ALDEP which needs to be modified further.

The basic concept with reference to ALDEP is explained with the help of Example 2.6.

Example 2.6

Relationship chart showing the closeness ratings between any two departments is shown below:

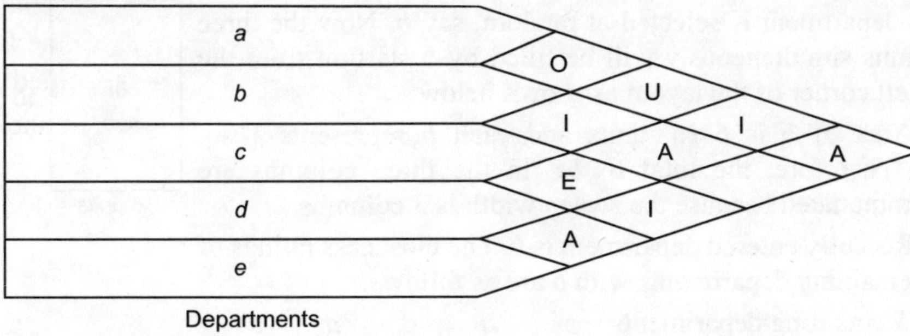

Five departments *a*, *b*, *c*, *d* and *e* are to be located in a layout.

From the relationship diagram, closeness rating between *a* and *b* is O, between *a* and *c* is U and so on. Obviously, rating between *b* and *a* will also be O and between *c* and *a*, it will be U.

Closeness rating for the pair of departments are summarized as follows:

Pair of departments	Closeness rating
a-b/b-a	O
a-c/c-a	U
a-d/d-a	I
a-e/e-a	A
b-c/c-b	I
b-d/d-b	A
b-e/e-b	O
c-d/d-c	E
c-e/e-c	I
d-e/e-d	A

Area of the departments are:

Department	:	*a*	*b*	*c*	*d*	*e*
Area in sq.m. (× 100)	:	9	6	15	18	12

Ignoring (× 100) sq.m. which is common to all departments, the total area of the departments/layout is 60 sq.m. Let us say that the total area available is (10×6) sq.m. as shown below:

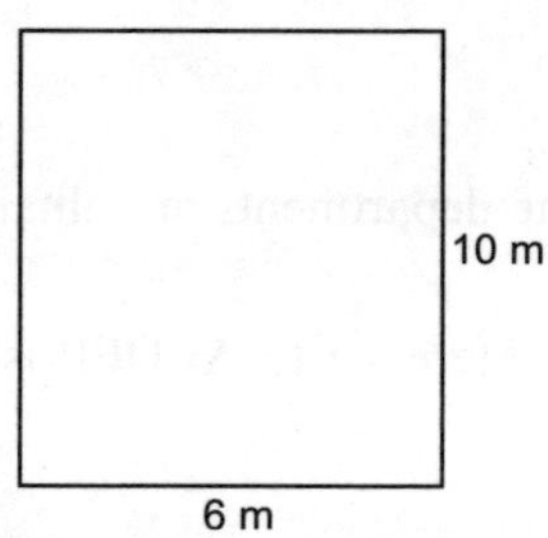

Assume that the sweep width = 3 columns.

Cycle 1

First department is selected at random, say *b*. Now the three columns simultaneously will be filled by *b* starting from the top-left corner of the layout as shown below:

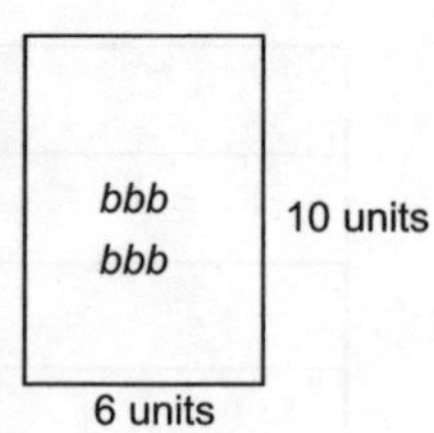

Area of *b* is 6 sq. units and each *b* represents 1 sq. unit. Therefore, the total 6 '*bs*' in the three columns are accommodated because the sweep width is 3 columns.

Recently entered department is *b*. The closeness ratings of the remaining departments with *b* are as follows:

Remaining departments	:	*a*	*c*	*d*	*e*
Rating with *b*	:	O	I	A	O

The highest rating is A corresponding to the department *d*, therefore *d* is selected next to enter in the layout.

[Note: If minimum department preference value is specified as, say I in terms of the closeness rating and if all ratings are lower than I, then any department would be selected at random.]

Area of *d* is 18 sq. units and therefore the six rows would be filled in three-column width as shown:

b	*b*	*b*
b	*b*	*b*
b	*b*	*b*
b	*b*	*b*
b	*b*	*b*
b	*b*	*b*
b	*b*	*b*
b	*b*	*b*

10 units

6 units

[Out of 10 units, in the first three-column width, 8 units are filled. Two rows are empty in this column width.]

Recently selected department is *d*.

Remaining departments : *a* *c* *e*

Rating with *d* : I E A

The highest rating is A corresponding to the department *e*, therefore, *e* is selected to enter. Area of *e* is 12 sq. units and it will require four rows of three-column width. After filling two rows in the first sweep width, the bottom of the layout will be reached. Now start from the bottom of the second sweep width of three columns as shown below:

b	*b*	*b*			
b	*b*	*b*			
d	*d*	*d*			
d	*d*	*d*			
d	*d*	*d*			
d	*d*	*d*			
d	*d*	*d*			
d	*d*	*d*			
e	*e*	*e*	*e*	*e*	*e*
e	*e*	*e*	*e*	*e*	*e*

10 units

6 units

Recently entered department is *e*.

Remaining departments : *a* *c*

Rating with *e* : A I

Department *a* is selected and filled in the layout as shown below:

b	*b*	*b*			
b	*b*	*b*			
d	*d*	*d*			
d	*d*	*d*			
d	*d*	*d*			
d	*d*	*d*	a	a	a
d	*d*	*d*	a	a	a
d	*d*	*d*	a	a	a
e	e	e	e	e	e
e	e	e	e	e	e

10 units

6 units

Remaining department is only *c*, it is entered as shown below:

b	*b*	*b*	*c*	*c*	*c*	
b	*b*	*b*	*c*	*c*	*c*	
d	*d*	*d*	*c*	*c*	*c*	
d	*d*	*d*	*c*	*c*	*c*	
d	*d*	*d*	*c*	*c*	*c*	10 units
d	*d*	*d*	*a*	*a*	*a*	
d	*d*	*d*	*a*	*a*	*a*	
d	*d*	*d*	*a*	*a*	*a*	
e	*e*	*e*	*e*	*e*	*e*	
e	*e*	*e*	*e*	*e*	*e*	

6 units

Fig. 2.21 Proposed layout at the end of cycle 1

Proposed layout as above by ALDEP may be imagined in the following form:

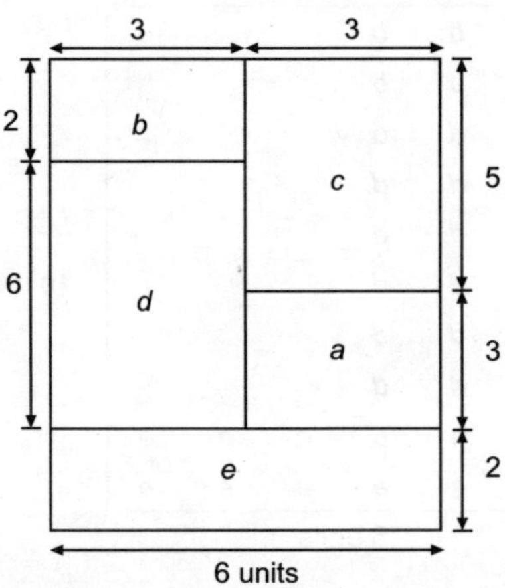

Pairs of the adjacent departments are *b-c*, *b-d*, *d-c*, *d-a*, *d-e*, *e-a* and *a-c*. The objective is to maximize the total closeness rating of the pairs of adjacent departments. Total closeness rating for the proposed layout at the end of cycle is obtained as follows:

Pair of Adjacent Departments	Closeness Rating	Value
b-c	I	4
b-d	A	64
d-c	E	16
d-a	I	4
d-e	A	64
e-a	A	64
a-c	U	0
	Total closeness rating =	**216**

The process may be repeated many times by considering any other department to enter first in the layout. Total rating is evaluated and the layout with maximum total rating is retained as the solution.

For example in the cycle 2, *c* may enter first.

Cycle 2

c is the first department in the sequence. Remaining departments are *a*, *b*, *d* and *e* and their rating with *c* are U, I, E, I, respectively. Maximum rating is E with respect to the department *d* and therefore *d* is selected.

Now the selected department is *d*.

Remaining departments:	*a*	*b*	*e*
Rating with *d*	: I	A	A

Both *b* and *e* have maximum rating. Let *e* be selected.

Remaining departments:	*a*	*b*
Rating with *e*	: A	O

Department *a* is selected and then finally *b* will enter. The sequence obtained is *c-d-e-a-b*. After filling the departments in this sequence, the layout will look like as shown below:

c	*c*	*c*	*b*	*b*	*b*
c	*c*	*c*	*b*	*b*	*b*
c	*c*	*c*	*a*	*a*	*a*
c	*c*	*c*	*a*	*a*	*a*
c	*c*	*c*	*a*	*a*	*a*
d	*d*	*d*	*e*	*e*	*e*
d	*d*	*d*	*e*	*e*	*e*
d	*d*	*d*	*e*	*e*	*e*
d	*d*	*d*	*e*	*e*	*e*
d	*d*	*d*	*d*	*d*	*d*

10 units

6 units

The above layout may be imagined as follows where *d* an odd shape.

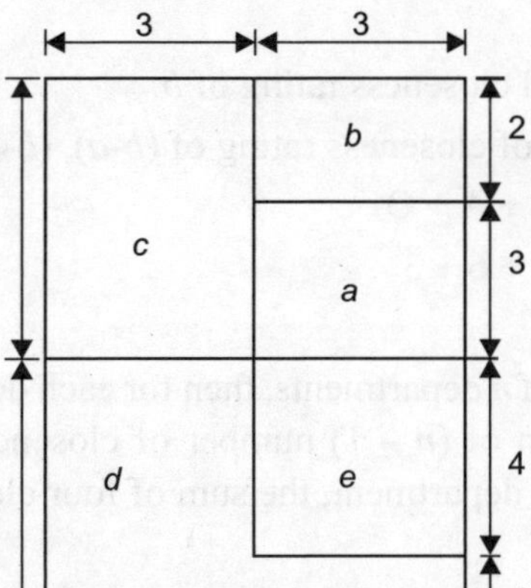

Pair of adjacent departments	Closeness Rating	Value
c-b	I	4
c-a	U	0
c-d	E	16
d-e	A	64
e-a	A	64
a-b	O	1
	Total closeness rating =	**149**

Total number of cycles to be computed may be specified. If only two cycles are to be performed, then the maximum value of total rating is 216 and Fig. 2.21 is retained as the proposed solution.

2.6.7 CORELAP (Computerized Relationship Layout Planning)

It is also a construction routine. The following closeness rating values are used:

Closeness rating	Value
A	6
E	5
I	4
O	3
U	2
X	1

First department is not selected at random, but the selection is based on the total closeness rating. Consider the relationship chart of the example 2.6.

Total closeness rating of department *a*

= Sum of closeness rating of a with the remaining departments

= Sum of closeness rating of (*a-b*), (*a-c*), (*a-d*), (*a-e*)

= O + U + I + A

= 3 + 2 + 4 + 6

= 15.

Similarly the total closeness rating of *b*

= Sum of closeness rating of (*b-a*), (*b-c*), (*b-d*) & (*b-e*)

= O + I + A + O

= 3 + 4 + 6 + 3

= 16.

If there are total of *n* departments, then for each department, the total closeness rating will be the sum of $(n - 1)$ number of closeness ratings. As there are five departments, for each department, the sum of four closeness ratings will yield the total closeness rating.

Department with the maximum total closeness rating is selected and placed at the centre of the layout. In case of a tie, the department with maximum area is selected first.

2.7 Formation of Cells in the Cellular Manufacturing Layout

As discussed in section 2.4(d), cellular manufacturing layout is suitable for the group technology approach. Usually for an existing job shop, some items are to be identified which can be produced in the medium quantities. The items are grouped based on their design requirements or operational requirements. Depending on the requirement of the group of items, necessary equipment are installed together in order to form a manufacturing cell.

2.7.1 Cell Formation

Consider that a job shop has identified 6 items for which there is a keen interest in knowing whether the manufacturing cells can be formed or not. A cell will comprise certain machines and each cell will act as a product layout for specified items because the machines are arranged according to the operations to be carried out. Specified items will be completely manufactured in the cell itself.

As an example, the machine-item details are shown in the following matrix:

		Item					
		1	2	3	4	5	6
Machine	1	1	0	0	1	0	1
	2	0	1	1	0	1	0
	3	1	0	1	1	0	1
	4	0	1	1	0	1	0
	5	1	0	0	1	0	1

Five types of machines are there which will be used as per the operations to be performed on items. Third row corresponding to machine 3 is 1 0 1 1 0 1, which means machine 3 will be required to produce item no. 1, 3, 4 & 6. In other words, items 1, 3, 4 and 6 will need machine 3 for their processing. '1' refers to the requirement of machines for items. '0' refers to the non-requirement. '0' appears for items 2 and 5 corresponding to machine 3. Machine 3 is not required to process items 2 and 5, or items 2 and 5 do not need machine 3 for manufacturing them.

Column Adjustment

Observe the columns. Items 1, 4 and 6 require similar machines (*viz.*, 1, 3 and 5). Similarly items 2 and 5 require similar machines.

Rearrange these columns to make them adjacent as shown below:

		Item					
		1	4	6	2	5	3
Machine	1	1	1	1	0	0	0
	2	0	0	0	1	1	1
	3	1	1	1	0	0	1
	4	0	0	0	1	1	1
	5	1	1	1	0	0	0

Row Adjustment

Observe the rows. Machines 1, 3 and 5 are required to produce the items (1, 4 and 6) completely. Make them adjacent. Similarly machines 2 and 4 are required to produce items 2 and 5 completely. Rearrange the rows as follows:

		Item					
		1	4	6	2	5	3
Machine	1	1	1	1	0	0	0
	3	1	1	1	0	0	1
	5	1	1	1	0	0	0
	2	0	0	0	1	1	1
	4	0	0	0	1	1	1

Now two cells can be formed. In one cell, machines 1, 3 and 5 and in the other cell, machines 2 and 4 can be installed as shown below:

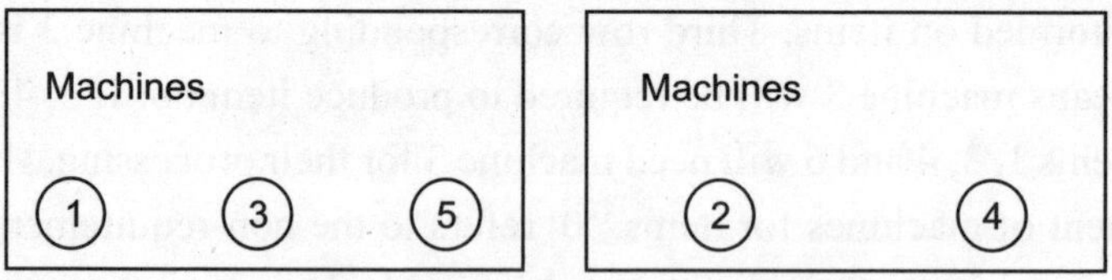

Cell 1 Cell 2

Cell 1 is devoted to the manufacture of items 1, 4 and 6.

Cell 2 is used to manufacture items 2 and 5.

Item 3 cannot be grouped with any of the cells because it requires the processing in both the cells. Machine 3 of the cell 1, as well as machines 2 and 4 of the cell 2, are required to manufacture item 3. Processing of item 3 may be treated as a job shop or functional layout. Material handling is more for item 3.

2.7.2 Bond Energy Algorithm

Cells are also formed with the use of bond energy algorithm. For example, if the machines–item details are as follows:

		Item			
		1	2	3	4
Machine	1	1	0	0	1
	2	0	1	1	0
	3	1	0	1	1
	4	0	1	1	0

then the bond energy between any two rows say 1–3 will be the sum of multiplication of the corresponding numbers of row 1 and row 3.

Row 1: 1 0 0 1

Row 3: 1 0 1 1

Bond energy between rows 1 and 3

$= (1 \times 1) + (0 \times 0) + (0 \times 1) + (1 \times 1)$

$= 1 + 0 + 0 + 1$

$= 2.$

Similarly bond energy between rows 1 and 4

$= (1 \times 0) + (0 \times 1) + (0 \times 1) + (1 \times 0)$

$= 0$

Following the same procedure, the bond energy between any two columns, say 1 and 3 will be the sum of multiplication of corresponding number in column 1 and column 3.

Column 1	Column 3
1	0
0	1
1	1
0	1

Bond energy between column 1 and column 3

$= (1 \times 0) + (0 \times 1) + (1 \times 1) + (0 \times 1)$

$= 0 + 0 + 1 + 0$

$= 1$

In this way, the bond energy may be computed for any two rows or for any two columns.

To apply the bond energy algorithm for the formation of cells, first of all, a column is selected at random. Bond energy of this column with remaining each column is computed and the column with maximum bond energy is chosen to be adjacent to the earlier randomly selected column. A separate matrix is formed by selecting one column after another following the procedure with the remaining columns. This recently obtained matrix is considered for the rows-wise adjustment now and finally an attempt is made to form suitable cells.

Example 2.7

Consider the following machine-item details:

		Item			
		1	2	3	4
Machine	1	0	1	1	0
	2	0	0	0	1
	3	1	0	0	1
	4	0	1	0	0
	5	1	0	0	1
	6	0	1	1	0

Now, column 2 is selected at random. In a separate matrix, column 2 will be arranged as the first column.

Remaining columns:	1	3	4
Bond energy of column with the remaining columns = Bond energy between the columns:	2-1	2-3	2-4
Bond energy:	0	2	0

Maximum bond energy is 2 corresponding to column 3, therefore column 3 is selected next to enter in the matrix.

As the selected column is 3,

Remaining columns :	1	4
Bond energy between the columns :	3-1	3-4
Bond energy	0	0

Both columns have equal bond energy with the column 3, select any column, say 1.

Remaining column is only column 4; it will be the last in the sequence.

Sequence of columns in the separately formed matrix is 2-3-1-4 as follows:

Machine \ Item	2	3	1	4
1	1	1	0	0
2	0	0	0	1
3	0	0	1	1
4	1	0	0	0
5	0	0	1	1
6	1	1	0	0

The above matrix will be treated as the reference matrix for a row-wise adjustment.

Select any row, say 4, randomly. The bond energy of row 4 with the remaining rows will be that of pair of rows 4-1, 4-2, 4-3, 4-5 and 4-6.

Remaining rows:	1	2	3	5	6
Bond energy with 4:	1	0	0	0	1

Both rows 1 and 6 have the maximum bond energy, *i.e.* 1 with row 4. Select any, say row 1.

Remaining rows:	2	3	5	6
Bond energy with row 1:	0	0	0	2

Maximum bond energy is 2 corresponding to row 6, therefore row 6 is selected next.

Remaining rows:	2	3	5
Bond energy with row 6:	0	0	0

Select any row, say 2.

Remaining rows:	3	5
Bond energy with row 2:	1	1

Select any row, say 3. The remaining row is row 5 only. The rows selected are in the sequence 4-1-6-2-3-5. The new matrix will be formed considering the adjustment of rows in this sequence as follows:

Machine \ Item	2	3	1	4
4	1	0	0	0
1	1	1	0	0
6	1	1	0	0
2	0	0	0	1
3	0	0	1	1
5	0	0	1	1

The two cells can be formed as follows:

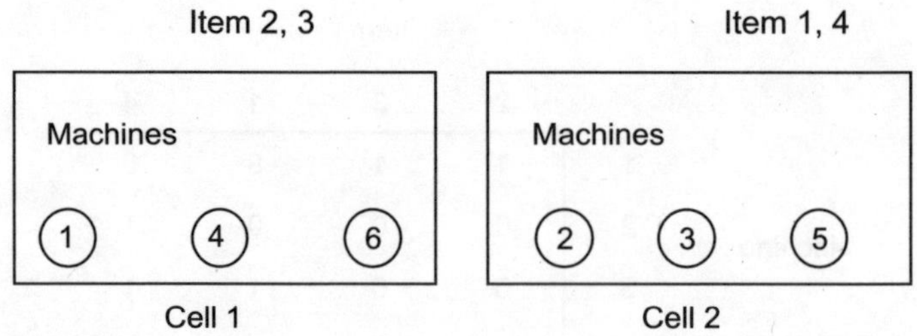

Machines 1, 4 and 6 are arranged in cell 1 in which items 2 and 3 are processed. Machines 2, 3 and 5 are installed in cell 2 for processing the items 1 and 4.

EXERCISES

1. Why is the proximity to raw materials and market a significant factor affecting the plant location?
2. What are the advantages of having ancillary units in the proximity of the main industry?
3. "The industries must be scattered throughout the nation from the strategic point of view." Comment.
4. Explain the following statement "Sub-urban area seems to be a compromise between the urban and rural regions".
5. Discuss the objective, subjective and critical factors affecting the selection of an industrial location.
6. What are the basic flow lines for material flow? Describe their characteristic features.
7. State whether the line layout is suitable for ship building. If not, what is the reason?
8. Differentiate between product and process layouts.
9. Under what circumstances, would the cellular manufacturing layout be suitable.
10. "Assembly line balancing is an important area of the analysis of product layout." Comment.
11. Tasks from A to J are to be performed in an assembly line. There is a requirement of 400 products per day. Assume the productive time as 8 hours per day.

The task time and the immediately preceding tasks for each task are as follows:

Tasks	Immediately Preceding Tasks	Task Time (seconds)
A	–	25
B	A	25
C	A	35
D	C	40
E	C	25
F	B	50
G	D, E	50
H	F	20
I	G, H	35
J	I	60

Balance the assembly line and obtain the efficiency of your proposal.

12. Discuss the incremental utilization heuristic briefly. Using this heuristic, assign the tasks to workstations for the following precedence diagram of tasks. Time taken by each task is also provided in minutes.

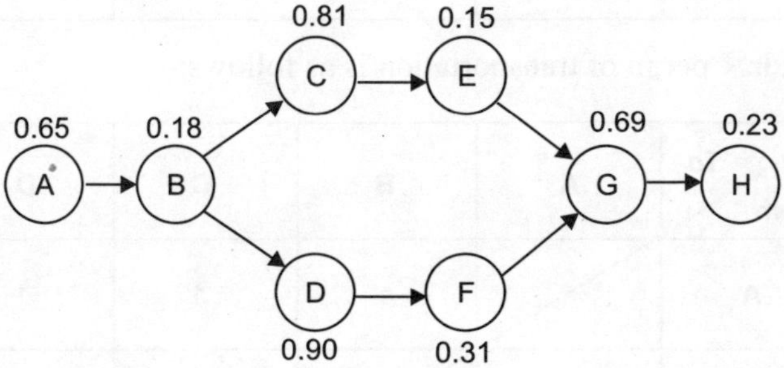

Assume that the assembly line will operate for 400 minutes per day and 2000 products are required to be assembled each day. Also evaluate the proposed solution by computing efficiency of the proposal.

13. Name the various techniques regarding planning for the process layout.

14. How is systematic layout planning carried out?

15. Discuss various assumptions in the CRAFT.

16. Implement the CRAFT for the following initial layout

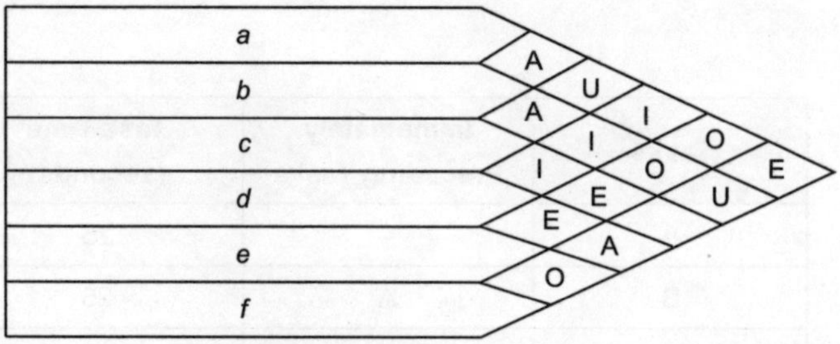

Load to be transported in a certain period is shown in the movement matrix as follows:

From \ To	A	B	C	D
A		2	1	3
B	3		4	2
C	2	3		4
D	0	0	1	

Cost in ₹ per m of transportation is as follows:

From \ To	A	B	C	D
A		1	1	1
B	2		1	1
C	2	1		1
D	1	1	3	

17. How is the COFAD different from the CRAFT?

18. Discuss various closeness ratings in the ALDEP.

19. Six departments, *a*, *b*, *c*, *d*, *e* and *f*, are to be located in a layout of equal length and width, *i.e.* having a square shape. The relationship chart is as follows:

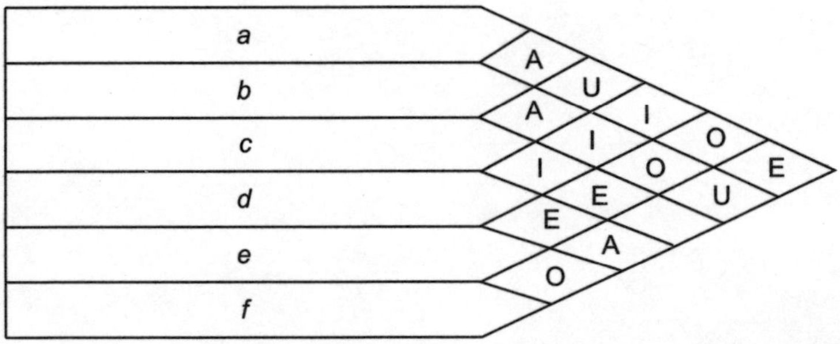

Departments

Area of each department is as follows:

Department	:	*a*	*b*	*c*	*d*	*e*	*f*
Area in sq.m (× 100)	:	6	18	20	6	20	30

Assuming the sweep width to be 2 columns, implement the ALDEP.

20. How is the CORELAP different from the ALDEP?

21. Implement the bond energy algorithm for the following machine-item details and comment whether the suitable cells can be formed or not.

Item

		1	2	3	4	5	6
Machine	1	1	0	1	0	1	1
	2	0	1	0	1	0	1
	3	0	1	0	1	0	1
	4	1	0	1	0	1	1
	5	1	0	1	0	1	0

❑❑❑

3

Demand Forecasting

Forecasting means assessment of the future. All individuals as well as organizations are interested in the knowledge of the future, *i.e.* assessment of the future periods. Demand forecasting is associated with predicting the demands of products or services. Consider the following products structure:

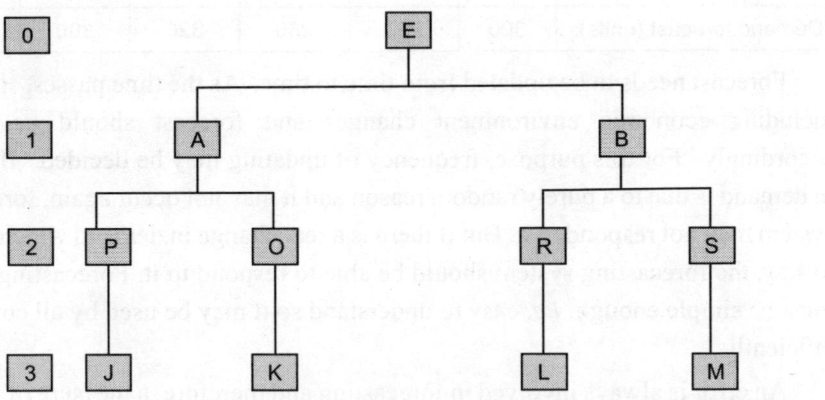

At the top, *i.e.* at '0' level there is the end product, E, which consists of two sub-assemblies A and B at level '1'. At level '2', P and Q are two components/parts required for sub-assembly A. Similarly R and S are two parts required to produce B. At level '3', J, K, L and M are raw materials needed to produce parts P, Q, R and S respectively. Parts S and P are also being marketed as spare parts in addition to manufacturing the end item, E.

For example, if the forecast of demand for the end item, E, during the next month is 300, then demand of the lower-level item may be calculated accordingly and there is no need to forecast the demand of lower-level items. Computation is based on their requirement per unit end item. But parts S and P are also being sold as spare parts. Their requirements to manufacture E can easily be computed, but fraction of their total demand which will be consumed as spare parts, need to be forecast. Thus, demands may be forecast or calculated, depending on the situations. Wherever possible, demands should be computed rather than forecast.

In the present example, if the requirement of raw material, K, is 15 kg per unit end item, then calculated demand of K = 15 × 300 = 4500 kg for the next month.

3.1 Development of Forecasting System

Accuracy level of the forecast should be decided for development of a suitable forecasting system. More accurate the forecast, more the efforts; and thus increased costs. A trade-off between accuracy and cost is necessary. Planning horizon is to be determined because short-range forecasts are more accurate than long-range forecasts. In the short-term forecasting (say one year or less than that), taste of the customers, market trends, etc., may behave as predicted. But in the long-term forecasting, uncertainty is very high.

Planning horizon is to be divided into suitable periods. If planning horizon is, say one year, then it may be divided into months/weeks/days as time period for forecasts. For example, the forecast of the product for next 6 months may be:

Month:	1	2	3	4	5	6
Demand forecast (units):	300	250	280	320	300	275

Forecast needs to be updated from time to time. As the time passes, situations including economic environment changes and forecast should be revised accordingly. For this purpose, frequency of updating may be decided. If change in demand is due to a purely random reason and it may not occur again, forecasting system may not respond to it. But if there is a real change in demand which is there to stay, the forecasting system should be able to respond to it. Forecasting system must be simple enough, *i.e.* easy to understand so it may be used by all concerned efficiently.

An error is always involved in forecasting and therefore, a measure of forecast error should be included.

3.2 Applications of Demand Forecasts

Directly or indirectly, demand forecasts are used for the following:

1. **Production Planning:** Depending on the demand forecast and production economy, production schedule is generated which will state the production quantity in respective periods.
2. **Capacity Planning:** Capacity may be in terms of machine capacity, human resources, storage space, etc. For example, as per the production schedule, whether machine capacity, storage capacity and human as well as other resources are available or not. If available capacity is less than the production plan, then a decision is to be taken whether extra capacity must be arranged or production plan should be altered in accordance with the available capacity.

3. **Profit Projection:** As per the forecast, sales revenue and total cost as well as profit may be projected for future periods and viability of the existing as well as planned investments may be verified.

3.3 Methods of Forecasting

Different methods of forecasting are shown in Fig. 3.1.

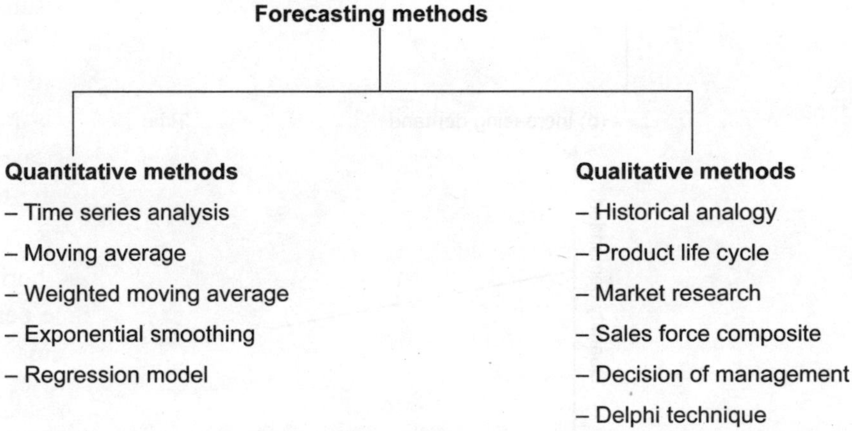

Fig. 3.1 Forecasting methods

3.3.1 Quantitative Methods

There are situations when sufficient data regarding the past or the history of the demand of the product is available. Quantitative methods are useful in analyzing such data and in forecasting of the demand. Following are the various quantitative methods:

(*a*) **Time Series Analysis:** The data related to actual demand are plotted against time and appropriate conclusions are drawn. Fig. 3.2 shows various patterns of demand. Fig. 3.2 (a), (b) and (c) show constant, increasing and decreasing demand patterns with respect to time. Accordingly, future demands may be projected extending the line further.

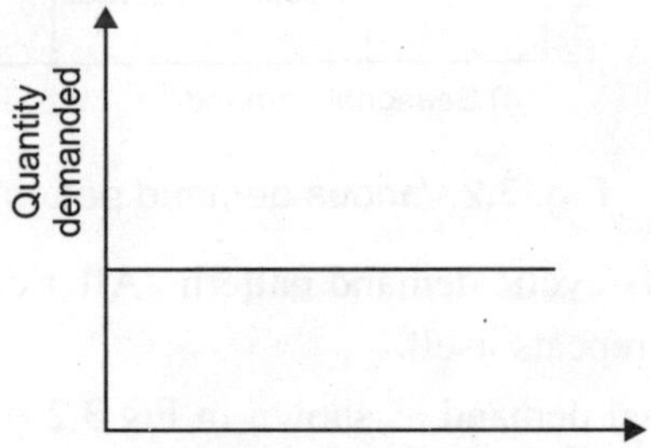

(a) Constant demand

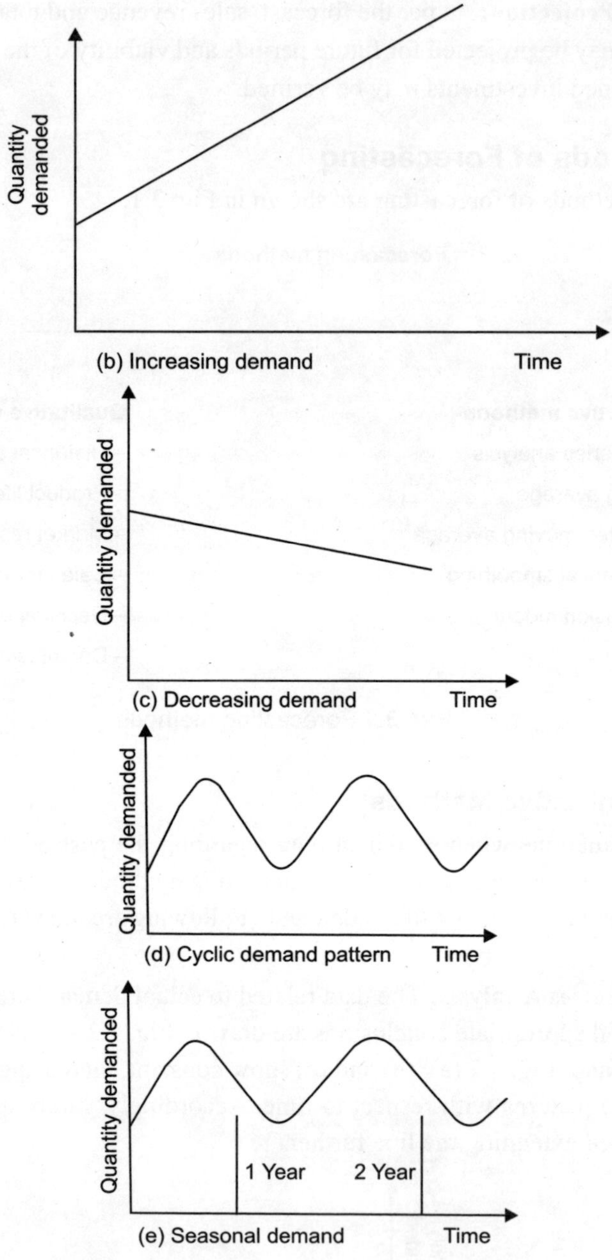

Fig. 3.2 Various demand patterns

Fig. 3.2 (d) shows cyclic demand pattern. After certain time (may be many years), the cycle repeats itself.

In case of seasonal demand as shown in Fig.3.2 (e), cycle completes in one year.

(b) **Moving Average:** In this method, the number of periods for which average of demands will be computeds is decided by the planners. For example, if four-period average is to be considered and if at the end of the current period i, forecast f_i for period $(i + 1)$ is to be made, then

$$f_i = (d_i + d_{i-1} + d_{i-2} + d_{i-3})/4$$

Where d_i, d_{i-1}, d_{i-2} & d_{i-3} are actual demands in period i, i–1, i–2 & i–3 and f_i = forecast made at the end of period i for next period $(i + 1)$. At the end of period $(i + 1)$, if actual demand occurred is d_{i+1}, then:

$$f_{i+1} = (d_{i+1} + d_i + d_{i-1} + d_{i-2})/4$$

In this way recent four periods will always be considered in this example for computing average and the most distant demand will be dropped for each successive forecast. For computing f_{i+2}, d_{i-2} is not needed and will be dropped. d_{i+2} will enter the calculation which is the most recent actual demand data:

Week:	1	2	3	4	5	6
Demand (Units):	5	4	6	7	7	5

Now considering four-period moving average, forecast at the end of week 6 for next, *i.e.* 7th week.

$$\begin{aligned} f_6 &= (d_6 + d_5 + d_4 + d_3)/4 \\ &= (5 + 7 + 7 + 6)/4 \\ &= 6.25 \approx 6. \end{aligned}$$

Forecast made in week 6 for next week is 6. But at the end of 7th week, actual demand is observed to be 7, then:

Week:	1	2	3	4	5	6	7
Demand (units):	5	4	6	7	7	5	7

Now

$$\begin{aligned} f_7 &= (d_7 + d_6 + d_5 + d_4)/4 \\ &= (7 + 5 + 7 + 7)/4 \\ &= 6.5 \approx 7. \end{aligned}$$

As the period for taking average, move with time, that is why the name "moving average" method of forecasting.

(*c*) **Weighted Moving Average:** Consider following demand data:

Week:	1	2	3	4	5	6
Demand (units):	3	5	4	2	1	3

Say, the weightage provided to the most recent data is 40%, then 30%, 20% and 10% to consecutive older data, i.e. again considering 4 weeks computation:

Week:	6	5	4	3
Demand:	3	1	2	4
Weightage (%):	40%	30%	20%	10%

(Sum of weightage is 100%)

Weighted moving average

$$= (3 \times 0.4) + (1 \times 0.3) + (2 \times 0.2) + 4\,(0.1)$$

$$= 2.3.$$

which is considered as forecast f_6 for 7th week. As discussed in case of moving average, at the end of 7th week, actual demand occurred in that week will be used in computation and the oldest data, *i.e.* for week 3 will be left out. In moving average, equal weightage was provided to all specified period data. But in the present method, weightage is provided depending on the prevailing situation. Sometimes older data is given more weightage than the recent data.

(*d*) Exponential Smoothing

Let d_i = Actual demand in period i,

f_{i-1} = Forecast made in period $(i-1)$ for period i,

f_i = Forecast made in period i for next period $(i + 1)$,

α = Smoothing constant (value between 1 and 0), and

$$f_i = \alpha\, d_i + (1 - \alpha) f_{i-1}.$$

If $\alpha = 0.25$, then $f_i = 0.25\, d_i + 0.75\, f_{i-1}$.

In other words, forecast made for the next period is the weighted average of current actual demand and previous forecast. Therefore, only previous forecast and current demand need to be stored and data storage required is less. If actual change in customer demand has occurred which is there to stay, then value of α is high and forecasting system is able to respond to real change in demand. At the same time, value of α chosen is small if change in customer demand is purely due to random reasons and it may not repeat itself, thereby not responding to purely random fluctuation in demand. In such a case, there is more reliance on the previous forecast.

Now $f_i = \alpha.\, d_i + (1 - \alpha) f_{i-1}$.

Substituting $f_{i-1} = \alpha.d_{i-1} + (1 - \alpha) f_{i-2}$

$$f_i = \alpha.d_i + (1 - \alpha)\,[(\alpha.d_{i-1} + (1 - \alpha) f_{i-2})]$$

Or $f_i = \alpha.d_i + \alpha(1 - \alpha)\, d_{i-1} + (1 - \alpha)^2 f_{i-2}$

Substituting $f_{i-2} = \alpha.d_{i-2} + (1 - \alpha) f_{i-3}$ in the above equation

$$f_i = \alpha.d_i + \alpha(1 - \alpha) d_{i+1} + (1 - \alpha)^2(\alpha.d_{i-2} + (1 - \alpha)\, f_{i-3}$$

$$= \alpha.d_i + \alpha(1 - \alpha) d_{i-1} + \alpha(1 - \alpha)^2 d_{i-2} + (1 - \alpha)^3 f_{i-3}$$

Continuing in this manner

$$f_i = \alpha d_i + \alpha(1 - \alpha) d_{i-1} + \alpha(1 - \alpha)^2 d_{i-2} + \alpha(1 - \alpha)^3\, d_{i-3} + \ldots$$

If $\alpha = 0.25$ say

$$f_i = 0.25 d_i + 0.1875\, d_{i-1} + 0.14 d_{i-2} + 0.105 d_{i-3} + \ldots$$

Maximum weightage is provided to the most recent demand data and then weightage declines exponentially with older demand data.

(*e*) **Regression Model:** If two variables are correlated, then with the help of regression model, suitable relationship is obtained between those variables.

Consider x as the advertisement expenditure and y as the sales revenue.

Independent variable = x.

Dependent variable = y.

Assume that there is a linear relationship between x and y, *i.e.* $y = a + bx$ as shown in Fig. 3.3.

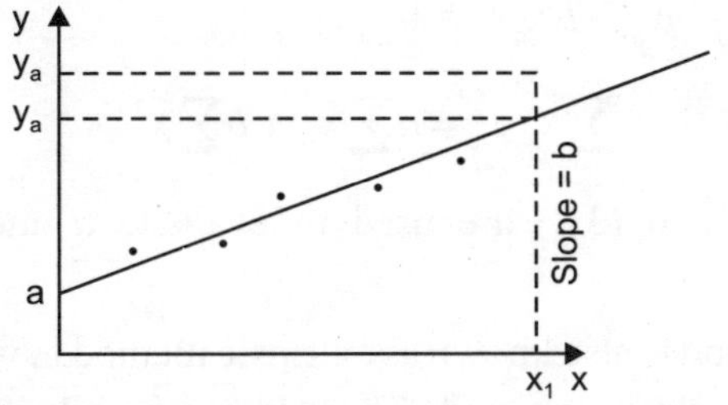

Fig 3.3 Relationship between two variables

a and b are constants which need to be obtained from the past available data.

Line $y = a + bx$ represents the forecasts for various values of x and dots in Fig.3.3 represents actual sales values of y for different x. At any value x_i, y_{ia} is the actual data whereas y_{if} is the forecast based on $y = a + bx$.

Deviation from actual = $y_{ia} - y_{if}$.

Mean square deviation is a measure of error which is:

$$\sum_{i=1}^{n} \frac{(y_{ia} - y_{if})^2}{n}$$

Where n = number of actual data available.

To minimize the error, $\sum_{i=1} (y_{ia} - y_{if})^2$ should be minimized

Say $$M = \sum (y_{ia} - y_{if})^2$$

Or $$M = \sum (y_{ia} - a - bx_{ia})^2$$

as $$y_{if} = a + bx_{ia}$$

(x_{ia}, y_{ia}) are available data for independent and dependent variables and i = 1,2,3….n.

To minimize M, differentiating it partially with respect to a and equating to zero,

$$\frac{\partial M}{\partial a} = \Sigma 2(y_{ia} - a - bx_{ia})(-x_{ia}) = 0$$

or $$\sum y_{ia} - \sum a - \sum bx_{ia} = 0$$

or $$\sum y_{ia} = na + b\sum x_{ia} \quad \text{... (3.1)}$$

Differentiating M partially with respect to b and equating to zero,

$$\frac{\partial M}{\partial b} = \Sigma 2(y_{ia} - a - bx_{ia})(-x_{ia})$$

or $$\Sigma(x_{ia}.y_{ia} - ax_{ia} - bx_{ia})^2 = 0$$

or $$\sum x_{ia}y_{ia} = a\sum x_{ia} + b\sum x^2_{ia} \quad \text{... (3.2)}$$

Equations (3.1) and (3.2) are used to evaluate a and b and then linear relationship is obtained.

Regression method is also known as extrinsic method in which external factors related to demand are also associated. These factors may be from outside or inside the organization. Number of sales persons and advertising expenditure are factors on which demand of the product may depend and these are the factors from inside the firm. On the other hand, number of houses built may be correlated to sales of the furniture. For the furniture manufacturing industry, number of houses built is a factor from outside the industry.

Remaining quantitative methods, *viz*. time series analysis, moving average, weighted moving average and exponential smoothing are intrinsic methods which rely solely on the history of demand of the product.

Example 3.1

A firm has following data available for advertisement expenditure and its sales revenue in ₹. Obtain the relationship between two variables. Next time, the firm is desirous of investing ₹ 9×10^4 for advertising its product. Estimate the sales revenue corresponding to this much advertising expenditure.

S. No.	Advertisement expenditure ₹ ($\times 10^4$)	Sales revenue ₹ ($\times 10^6$)
1	3	14
2	4	15
3	3	11
4	5	18
5	4	12
6	6	20
7	7	26
8	4	13

Equations (3.1) and (3.2) are used to determine a and b, after obtaining a and b, relationship is as follows:-

$$Y_{if} = a + bx_{ia} \quad \text{...(3.3)}$$

Now $n = 8$

And $\sum x, \sum y, \sum xy$ and $\sum x^2$ are computed as shown below:

x_{ia}	y_{ia}	$x_{ia} * y_{ia}$	x^2_{ia}
3	14	42	9
4	15	60	16
3	11	33	95
5	18	90	25
4	12	48	16
6	20	120	36
7	26	182	49
4	13	52	16
$\sum x_{ia} = 36$	$\sum y_{ia} = 129$	$\sum x_{ia} y_{ia} = 627$	$\sum x^2_{ia} = 176$

Substituting the values of $\sum x, \sum y, \sum xy$ and $\sum x^2$ in equations (3.1) and (3.2)

$$129 = 8a + 36b$$

Or $8a + 36b = 129$ ...(3.4)

and $627 = 36a + 176b$

or $36a + 176b = 627$

Solving (3.4) and (3.5)

$$a = 1.18$$

and $b = 3.32$

and the relationship is obtained as,

$$y_{if} = 1.18 + 3.32x_{ia}$$

If $x_{ia} = 9$

$$Y_{if} = 1.18 + (3.32 \times 9)$$

$$= 31.06$$

Therefore, for advertisement expenditure of ₹ 9×10^4, the forecast is made for sales revenue and estimated sales revenue is ₹ 31.06×10^6.

Example 3.2

Demand of a product is growing in the following way:

Year	Demand in numbers (×1000)
2007	9
2008	11
2009	12
2010	15
2011	16
2012	19
2013	20

Forecast the demand of the product in 2014.

Now $n = 7$.

It becomes convenient to do scaling of the year as follows:

Year	x_{ia} = Year - 2010	y_{ia}	$x_{ia}\, y_{ia}$	X^2_{ia}
2007	–3	9	–27	9
2008	–2	11	–22	4
2009	–1	12	–12	1
2010	0	15	0	0
2011	1	16	16	1
2012	2	19	38	4
2013	3	20	60	9
	$\Sigma x_{ia} = 0$	$\Sigma y_{ia} = 102$	$\Sigma x_{ia}\, y_{ia} = 53$	$\Sigma x^2_{ia} = 28$

Using equations (3.1) and (3.2)

$$7a + 0 = 102$$

$$a = 102/7 = 14.57$$

and $\quad 28b = 53$

or $\quad b = 1.89$

Relationship is $\quad y_{if} = 14.57 + 1.89\, x_{ia}$

For Year 2014, $\quad x_{ia} = 2014 - 2010 = 4$

And $\quad y_{ia} = 14.57 + (1.89 \times 4) = 22.13.$

In the year 2014, the forecast of the demand of the product is 22.13 × 1000 = 22, 130.

Correlation Coefficient

Correlation coefficient, r between two variables, x and y is given by:

$$r = \frac{\text{Covariance}}{\sqrt{Variance(x)}\sqrt{Variance(y)}} = \frac{S_{xy}}{S_x S_y}$$

Where variance (x),

$$Sx^2 = \frac{\Sigma(x - x^-)^2}{n} \text{ and } Sx = \sqrt{\Sigma\frac{(x - x^-)^2}{n}}$$

Where x^- is the average of data x.

Similarly variance (y),

$$Sy^2 = \Sigma\frac{(y - y^-)^2}{n} \text{ and } S_y = \sqrt{\frac{\Sigma(y - y^-)^2}{n}}$$

Where y^- is the average of data y.

Covariance, $S_{xy} = \sum\frac{(x - x^-)(y - y^-)}{n}$

and
$$r = \Sigma\frac{(x - x^-)^2}{\sqrt{\Sigma(x - x^-)^2}}\frac{(y - y^-)^2}{\sqrt{\Sigma(y - y^-)^2}}$$

If correlation coefficient, r = –1, then the two variables are inversely related. If $r = 1$, then the variables are directly related.

$r = 0$ indicates that there is no correlation.

Level of correlation	Approximate range
Very less correlation	0–0.2
Less correlation	0.2–0.4
Medium correlation	0.4–0.7
High correlation	0.7–0.98
Very high correlation	0.98–1

Example 3.3

A tyre manufacturing industry wishes to correlate its sales of tyres with total number of cars manufactured and marketed in the country. In the past 10 years, data is available for each year which is as follows:

Year No.	Number of cars ($\times 10^6$)	Number of tyres sold ($\times 10^6$)
1	4	1
2	8	3
3	6	2
4	10	4
5	9	3
6	10	3
7	12	5
8	11	3
9	14	6
10	13	6

Find out if there is any correlation.

As the number of tyres sold may have dependence on number of cars arrived in the market, number of tyres is the dependent variable y. Number of cars is the independent variable x. Following table shows the computation:

x	y	$(x-\bar{x})$	$(x-\bar{x})^2$	$(y-\bar{y})$	$(y-\bar{y})^2$	$(x-\bar{x})(y-\bar{y})$
(1)	(2)	(3)	(4)	(5)	(6)	(3×5)
4	1	–5.7	32.49	–2.6	6.76	14.82
8	3	–1.7	2.89	–0.6	0.36	1.02
6	2	–3.7	13.69	–1.6	2.56	5.92
10	4	0.3	0.09	0.4	0.16	0.12
9	3	–0.7	0.49	–0.6	0.36	0.42
10	3	0.3	0.09	–0.6	0.36	0.18
12	5	2.3	5.29	1.4	1.96	3.22
11	3	1.3	1.69	–0.6	0.36	-0.78
14	6	4.3	18.49	2.4	5.76	10.32
13	6	3.3	10.89	2.4	5.76	7.92
$\Sigma x = 97$	$\Sigma y = 36$	$\Sigma(x-\bar{x})$	$\Sigma(x-\bar{x})^2$	$\Sigma(y-\bar{y})$	$\Sigma(y-\bar{y})^2$	$\Sigma(x-\bar{x})(y-\bar{y})$
		= 0	**= 86.1**	**= 0**	**= 24.4**	**= 42.8**

$$\bar{x} = \frac{97}{10} = 9.7 \qquad \bar{y} = \frac{36}{10} = 3.6$$

As correlation coefficient, $$r = \frac{\sum(x-\bar{x})}{\sqrt{\sum(x-x)^2}}\frac{(y-\bar{y})}{\sqrt{\sum(y-\bar{y})^2}}$$

$$= \frac{42.8}{\sqrt{86.1}\,\sqrt{24.4}}$$

$$= 0.93.$$

As $r = 0.93$, there is high correlationship.

Tracking Signal

Forecasting system should respond to the real change in demand either on higher or on lower side. Organization is always interested in knowing whether forecast is keeping pace with either increasing or decreasing change in demand. Tracking signal is a measure which indicates this. Tracking signal is the ratio of running sum of forecast errors (RSFE) and mean absolute deviation (MAD).

Example 3.4

An industry is using moving average method (with three-period averages) of forecasting.

Forecast and actual demand are given below:

Period:	1	2	3	4
Forecast:	5	6	7	6
Actual demand:	7	7	5	7

Calculations for RESF, MAD and tracking signal, i.e. RSFE/MAD are shown below:

Period	1	2	3	4
(1) Forecast	5	6	7	6
(2) Actual demand	7	7	5	7
(3) Deviation = (2 – 1)	2	1	-2	1
(4) RSFE (cumulative values considering (3))	2	3	1	2
(5) Absolute deviation (Abs. value for (3))	2	1	2	1
(6) Sum of AD (cumulative value considering (5))	2	3	5	6
(7) MAD ((6) + Period)	2	1.5	1.67	1.5
(8) Tracking signal ((4) + (7))	1	2	0.6	1.33

Tracking signals may be plotteds against period as shown below:

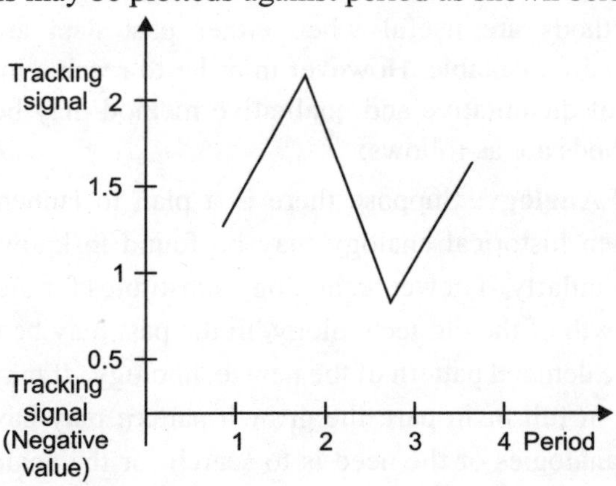

Tracking signal may be negative.

Example 3.5

Find out the standard error of estimate for the data of Example 3.1.

The relationship obtained was:

$$y_{if} = 1.18 + 3.32\, x_{ia}$$

Now $n = 8$ and for each value of x_{ia}, y_{if} is obtained as shown below:

x_{ia}	y_{ia}	y_{if}=1.18+3.32 x_{ia}	$(y_{ia} - y_{if})$	$(y_{ia} - y_{if})^2$
3	14	9.96	4.04	16.32
4	15	14.46	0.54	0.29
3	11	9.96	1.04	1.08

5	18	17.78	0.22	0.05
4	12	14.46	–2.46	6.05
6	20	21.1	–1.1	1.21
7	26	24.42	1.58	2.50
4	13	14.46	–1.46	2.13
				$\Sigma(y_{ia}-y_{if})^2$
				= 29.63

The standard error of estimate is given by:

$$\sqrt{\frac{\Sigma(y_{ia}-y_{if})^2}{n-2}}.$$

As $\Sigma(y_{ia}-y_{if})^2 = 29.63,$

and $n = 8.$

$$\text{Standard error of estimate} = \frac{\sqrt{29.63}}{6} = 2.22.$$

3.3.2 Qualitative Methods

Qualitative methods are useful when either past data are not available or insufficient data are available. However in order to reach a judicious conclusion, a combination of quantitative and qualitative method may be applied. Different qualitative methods are as follows:

(*a*) **Historical Analogy:** Suppose there is a plan to launch a new product or service, then historical analogy may be found to know the growth of the demand. Similarly, if newer technology substitutes for older technology, then rate of growth of the old technology in the past may be treated as a basis to forecast the demand pattern of the new technology. It may safely be assumed that either in full or in part, the growth pattern may have a bearing on the historical analogies or the need is to search for the analogous products and their history of demand. For example:

Product for which forecast is desired	Analogous product
Mobile telephones	Landline telephones
Colour T.V.	Black & White T.V.
Powered two wheelers	Bicycles
C.N.G.	Diesel

Black & White T.V. owners will sooner or later switch over to coloured T.V. sets. Similarly landline telephone consumers may have a trend towards shifting to use of mobile telephones.

(*b*) **Product Life Cycle:** Product life cycle is shown in Fig. 3.4.

This is a relationship between demand of a product and time.

Four zones have been marked on the curve:

I – Introduction
G – Growth
M – Maturity
D – Decline

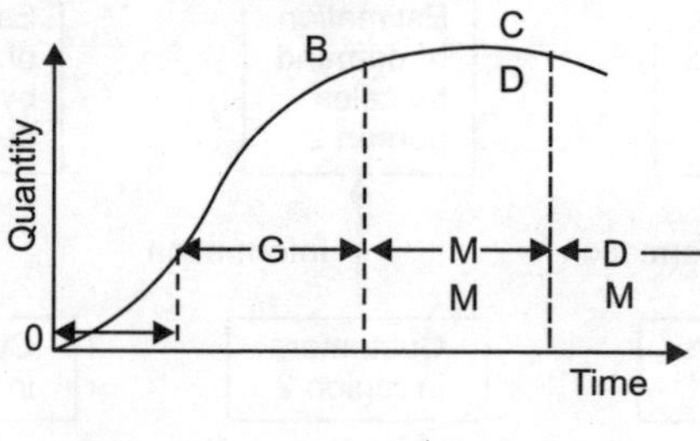

Fig. 3.4 Product life cycle

In the introduction region, the product is initially launched in the market and its demand rises from O to A at a slow rate. In the growth region, demand of the product increases at a faster rate and increases from A to B.

In the maturity zone, *i.e.* (B–C), the product becomes mature and demand is at its peak and almost constant.

Now as every product has its life span, from C onwards, the demand of the product starts declining.

(*c*) Market Research: Market research is a method by which the perception of the customer about a product is known.

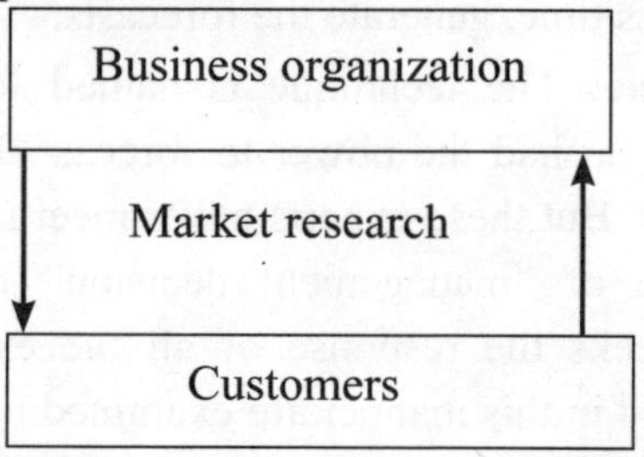

In the form of well-designed questionnaires, information is obtained from existing as well as potential customers. After data collection, the facts are analyzed and conclusions are drawn. After knowing the liking/disliking as well as by incorporating the suggestions for improvement, demand of the product may be created or enhanced. Statistical methods are used to analyze the data. Market research findings are useful in the demand forecasting.

(*d*) Sales Force Composite: Sales persons represent various regions in the country and they are in close contact with the customers. Depending on the customers' requirement in future, each sales person estimates the demand of the respective region. Then, these estimates are consolidated in order to develop the total demand forecast.

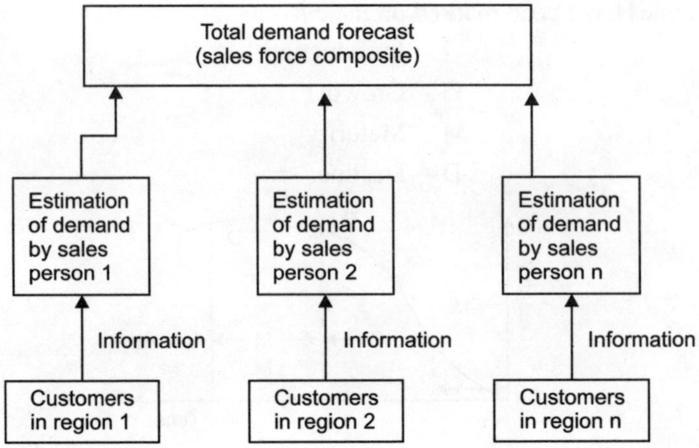

(*e*) **Decision of Management:** A committee may be formed by the management to forecast the demand. The committee may comprise experts from various disciplines such as product, marketing and financial management. Each person is free to express the opinion in the matters related to forecast. The job of the experienced chairperson of the committee is to allow a free discussion and then to draw an appropriate conclusion. The problem may occur when a powerful senior member of the committee gives his opinion and junior members either do not have courage to express a different view or are not allowed to share it. Care is to be taken to avoid such situations.

The objective of the method is to obtain different views and then in comparatively less time, generate the forecasts.

(*f*) **Delphi Technique:** The technique is named for the Delphic Oracle of ancient Greece who had the power to forecast. In this method, a panel of experts is formed. But these experts never meet in person, thereby avoiding the disadvantage of "management decision" method discussed before. A coordinator seeks the response of all the experts by correspondence. Forecasts received in this manner are examined by the coordinator. If all the forecasts are in closer range, then with less effort, an appropriate decision is made. Usually forecasts will be in a wide range and these are to be brought in a narrow range by repeated queries.

Consider a hypothetical example of steel-making industry. A panel of seven members is formed. Some of them are from inside the organization and some are experts from outside. An experienced coordinator is given the responsibility of execution of the task. The coordinator sends the designed performance of all seven members. The forecast is desired from each along with reasons to substantiate their level of forecast.

Forecasts obtained are as follows:

Member	Forecast (×10^5 M.T.)
1	2.92
2	1.90
3	3.08
4	3.05
5	2.95
6	4.00
7	3.00

After scrutiny of the forecasts received, the coordinator compile all the major reasons/assumptions forwarded by the members of the panel. It is also observed that member 2 has made the forecast on an exceptionally lower side. Forecast of member 6 is exceptionally high. Reason advanced by member 2 is as follows:

"A competitors is in the process of major expansion and due to that market share may go down."

This reason, along with other reasons, compiled by the coordinator is sent to member 6 and it is asked whether in the light of new information (particularly expansion of the competitor), a revision in the forecast is possible. Forecast made by member 6 was 4 × 10^5 M.T. earlier. Now with the recent information gained, revised forecast by member 6 is 3 × 10^5 M.T. and it is communicated by him to the coordinator.

The reason advanced by member 6 with his earlier forecast, *i.e.* 4 × 10^5 M.T. was "Due to government policy and change in economic scenario, it has become easier to export the steel and demand may go high."

This, as well as other reasons, are communicated to member 2 whose present forecast is 1.90 M.T. and a query is made whether there is any possibility of upward revision particularly due to export potential of the industry. The new information is taken into account by member 2 and revised forecast by him is now 2.97 M.T. Remaining members have also considered both the reasons, but they are of the opinion that decrease due to one reason and increase due to other reason are similar and therefore, their forecast is same as before. Now the forecasts are:

Member	Forecast (× 10^5 M.T)
1	2.92
2	2.97
3	3.08
4	3.05
5	2.95
6	3.00
7	3.00

The coordinator analyze the forecast and finds that these are in the closer range of (2.92–3.08), *i.e.* minimum forecast is 2.92×10^5 M.T. and maximum is 3.08×10^5 M.T. He finally concluded that forecast is approximately 3×10^5 M.T. In real practice, the process will be repeated several times in order to create a consensus. As the panel members never meet in person regarding the purpose of forecasts, they do not hesitate in revising their forecast after getting any new information. In this technique, possibility of compromise (as may occur in case of management decision discussed before) is completely eliminated and a true consensus is evolved. But this method takes a lot of time due to repeated correspondence and analysis of data.

3.4 Input–output Analysis

In this analysis, different sectors in the economy are included and aggregate input as well as output of the sectors are considered for overall planning. There are several sectors and their subdivisions in the economy of the country such as:

Agriculture	
Industry	Automobile
	Steel
	Textile
	Power
Services	I.T.
	Banking
	Education
	Insurance
	Telecommunication

But in the present hypothetical example, only automobile and I.T. have been considered. Output of the automobile sector is say, ₹ 3,000 crores and that of I.T. sector is ₹ 7,000 crore. Following from and to matrix shows how output of these sectors is consumed.

Data in thousands of crore of ₹

To →	Auto	I.T.	Remaining customers	Total
From ↓				
Auto	0.5	1	1.5	3
I.T.	2	2	3	7

Total output of auto sector is 3 out of which 0.5 goes to auto sector itself in the form of input, 1 to I.T. sector as input and the remaining 1.5 is consumed by variety of customers. Similarly, total output 7 of I.T. sector is distributed as follows:

2 to autos sector as input, 2 to I.T. sector as input, and remaining 3 is consumed by variety of customers.

As only two sectors are being considered, examine the first two columns to calculate the input–output coefficients. Total output of auto sector is 3 where 0.5 and 2 are inputs to this sector

Coefficients are $\frac{0.5}{3}$ and $\frac{2}{3}$.

Similarly total output of I.T. is 7 whereas inputs to it are 1 and 2 and coefficients are $\frac{1}{7}$ and $\frac{2}{7}$.

	Input–output	Coefficients
	To Auto	I.T.
From Auto	0.5/3	1/7
I.T.	2/3	2/7

If total output of auto is A and that of I.T. is I, then:

$$A-1.5 = \left[\frac{0.5}{3}\right]A+\left[\frac{1}{7}\right]I$$

Or $$\left[\frac{2.5}{3}\right]A-\left[\frac{1}{7}\right]I=1.5 \qquad \text{...(3.6)}$$

Similarly,

$$I-3 = \left[\frac{2}{3}\right]A+\left[\frac{2}{7}\right]I$$

Or $$\left[\frac{5}{7}\right]I-\left[\frac{2}{3}\right]A=3 \qquad \text{...(3.7)}$$

As the R.H.S. of Eqs. (3.6) and (3.7) is 1.5 and 3 respectively which are present consumption of variety of customer from auto & I.T. sectors, after solving (3.6) & (3.7), A and I may be obtained as 3 & 7 respectively which are total of these sectors.

If the customer demands for the next period are d_A and d_I for auto.& I.T. sectors, equations (3.6) & (3.7) may be written as:

$$\left[\frac{2.5}{3}\right]A-\left[\frac{1}{7}\right]I=d_A \qquad \text{...(3.8)}$$

$$\left[\frac{5}{7}\right]I-\left[\frac{2}{3}\right]A=d_I \qquad \text{...(3.9)}$$

If for any future period, d_A and d_I are projected to have certain values, equations (3.8) and (3.9) may be used to evaluate A and I which will be planned output of these sectors for future period. Accordingly a plan may be made for resources and other policy decisions.

For example, if $d_A = 2$ and $d_I = 3.5$ in the future period, then:

$$\left[\frac{2.3}{3}\right]A - \left[\frac{1}{7}\right]I = 2$$

And $$\left[\frac{5}{7}\right]I - \left[\frac{2}{3}\right]A = 3.5. \qquad \ldots(3.10)$$

Multiplying equation (3.10) by 5 and adding with (3.10)

$$\left[\frac{12.5}{3}\right]A - \left[\frac{5}{7}\right]I = 10$$

$$\left[\frac{2}{3}\right]A + \left[\frac{5}{7}\right]I = 3.5$$

$$\left[\frac{10.5}{3}\right]A = 13.5$$

Or $$A = \frac{40.5}{10.5} = 3.86.$$

Substituting Value of A in equation (3.11):

$$\left[\frac{5}{7}\right]I - \left[\frac{81}{31.5}\right] = 3.5$$

Or $$I = \frac{1338.75}{157.5} = 8.5.$$

Therefore, the approximate output requirement for auto & I.T. sectors will be 3.86×10^3 crore and ₹ 8.5×10^3 crore respectively for the future periods in order to satisfy the increased customer demands.

3.5 Relationship between Demand and Supply

Economic environment consists of production/supply as well as consumption/demand of goods and services. Both demand and supply have high dependence on price of product. Following three laws relate to it:

(*a*) Law of Demand

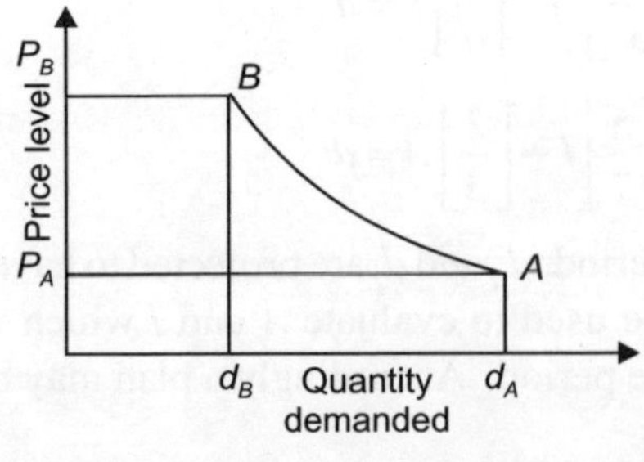

As shown in the above fig., at two points A and B, price of the product is P_A and P_B and the demand drops from d_A to d_B considering other similar factors:

$$\text{Proportionate change in demand} = \frac{(d_A - d_B)}{d_A}.$$

$$\text{Proportionate change in Price} = \frac{(P_B - P_A)}{P_A}$$

Price elasticity of demand is the ratio of proportionate decrease in demand and proportionate increase in price:

$$\text{Price elasticity of demand} = \frac{(d_B - d_A)/d_A}{(P_B - P_A)/P_A}$$

(*b*) Law of Supply

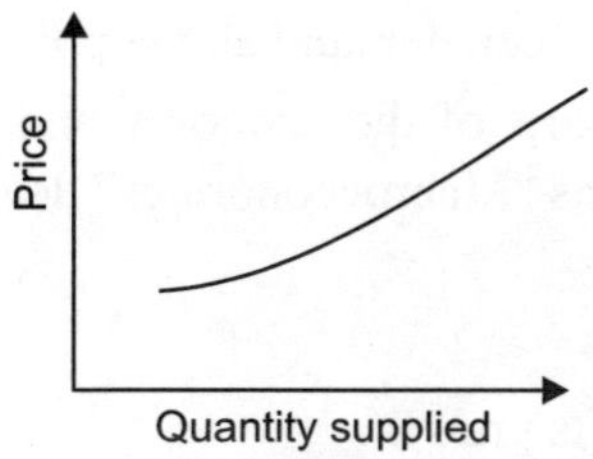

Law of demand was from customer side in which if price is more, demand decreases. But this law of supply is from producer or supplier side. If price of a commodity increases, producers and suppliers are encouraged to produce and supply more, therefore the supply tends to increases as the price increases.

(*c*) Law of Demand and Supply

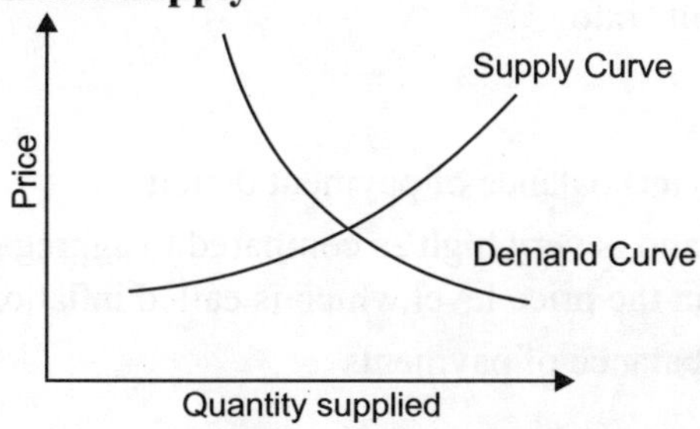

An equilibrium will be achieved with the combination of supply curve and demand curve. This is known as market price equilibrium. Price corresponding to the intersection of both the curves is called market price.

In fire market economy, there is inherent mechanism of balance of demand and supply as shown below:

Demand of the product goes up

↓ Results into

Shortage
(*i.e.* Demand > Supply)

↓ Results into

Price increase

↓ Results into

Supply increased
and demand decreased

↓ Until

Balance is created between demand and supply (*i.e.* Demand = Supply).

Study of individual parts of the economy and firms/industries comes under "Microeconomics" whereas "Macroeconomics" deals with aggregate demand and aggregate supply.

Macroeconomics Studies

(*a*) National output and its growth

(*b*) Balance of aggregate demand and aggregate supply

If there is an imbalance between aggregate demand and aggregate supply, then after certain effects are observed which may be as shown below:

Aggregate demand >> Aggregate supply

↓ Results into

(*a*) Inflation

(*b*) (Current account) balance of payment deficit

If aggregate demand is very high as compared to aggregate supply, then there is a general increase in the price level which is called inflation. Now:

Current account balance of payments

↓ Indicates

(Exports–Imports) of goods and services.

If exports are more than imports, *i.e.* (Exports-Imports) is a positive number, then it is called "current account surplus". If imports are more than exports, *i.e.* (Exports-Imports) is a negative figure, then it is called "current account deficit".

As the prices are rising due to aggregate demand being very high, more imports may be the natural reaction and "current account deficit" results.

But in case of reverse situation, *i.e.*

Aggregate demand << Aggregate supply

↓ Results into

(*a*) Recession

(*b*) Unemployment

If aggregate demand is very low in comparison with aggregate supply, then output in the economy declines. In other words, there is negative growth. This is called "Recession". As the output in the economy declines, unemployment is another end effect. Therefore, recession and unemployment take place when aggregate demand is too low relative to aggregate supply.

This overall understanding of the economic as well as business environment, directly or indirectly, helps in demand forecasting.

EXERCISES

1. What do you understand by demand forecasting? Discuss various applications of demand forecasts.
2. Differentiate between qualitative and quantitative methods of demand forecasting.
3. Write the difference between cyclic and seasonal demand patterns.
4. In order to implement moving average method, it was decided to consider six-period averages. Following data shows the actual demands in respective periods:

Periods	Actual Demand
1	6
2	4
3	5
4	7
5	4
6	7
7	8
8	6
9	5
10	9
11	7
12	8

Make forecasts for period 7 onwards and compare with the actual demands occurred.

In Exercise 4, consider the following weightage:

Most recent ….oldest data

25%, 22%, 20%, 15%, 10%, 8%

Apply weighted moving average method.

6. What are the advantages of exponential smoothing? Discuss the role of smoothing constant in this method.

Following data are given:

Current demand = 65

Previous forecast = 58

Smoothing constant = 0.2

Obtain the forecast for the next period using exponential smoothing.

7. A spark plug manufacturing industry wants to correlate its sales with the number of two-wheelers arrived in the market. The industry has the data available for the past 12 periods:

Periods	Number of two-wheelers ($\times 10^5$)	Number of spark plugs sold ($\times 10^5$)
1	5	2
2	9	4
3	6	2
4	12	6
5	10	4
6	9	3
7	11	5
8	13	4
9	14	6
10	13	5
11	9	3
12	8	3

(*a*) Find out if there is any correlation.

(*b*) Obtain the relationship between two variables.

(*c*) In the next period, if the number of two-wheelers to arrive in the market are 15×10^5, estimate the sales of spark plugs.

(*d*) Find out the standard error of estimate.

8. Demand of an item is growing as follows:

Year	Demand in numbers ($\times 10^4$)
2006	12
2007	14
2008	16
2009	19
2010	21
2011	22
2012	25
2013	26

Forecast the demand of the product in the year 2014.

9. What do you understand by tracking signal?

10. Suggest historical for the following products:

(*a*) Automatic washing machine

(*b*) Tape recorder

11. With reference to product life cycle, what measures may be taken before a decline in the demand of the product occurs.

12. What problems do you anticipate in market research?

13. With reference to sales force composite, sometimes sales persons intentionally estimate the demands on:

(*a*) Higher side

(*b*) Lower side

In each situation, what do they have in their minds?

14. Why, sometimes, is there a compromise in management decisions?

15. How is consensus evolved in Delphi technique? What are its advantages over Management Decision method?

16. "Input–output analysis" is useful in national planning. Comment.

17. Consider the following three sectors in the economy:

(*a*) Agriculture

(*b*) Industry

(*c*) Services

"From-to" details are as follows ($\times 10^3$ crores):

To:	Agriculture	Industry	Services	Remaining Consumers
From: Agriculture	4	1	1	1
Industry	0.5	2	1	1.5
Services	1.3	1.5	0.5	2

Conduct the input–output analysis if in the next period, consumer demands are:

₹ 1.1×10^3 crore for agriculture sector,

₹ 1.7×10^3 crore for industry sector,

and ₹ 2.1×10^3 crore for services sector.

Evaluate the total output for each these three sectors.

18. Explain the following:

(*a*) Law of demand

(*b*) Law of supply

(*c*) Law of demand and supply

19. The price of a product is ₹ 6 and the current demand is 500 numbers. Due to some reasons, there is a price rise for the product and now it is ₹ 6.50. But the demand is reduced to 460 numbers. Calculate the price elasticity of demand.

20. What are after effects if aggregate demand is too high relative to aggregate supply?

21. How is the understanding of economic environment useful in business decisions including demand forecasting?

❑❑❑

4

Production Planning and Control

Production planning and control has two components:

1. **Production Planning:** The planning for production–related activities are made at aggregate level as well as at the shop floor level.
2. **Production Control:** The production activities during execution are supervised closely. If there is any deviation from the planned targets, suitable action is taken either to eliminate the deviation or to minimize it.

4.1 Production Planning at Aggregate Level

Overall planning is done for producing specific goods in specific quantities with due consideration of profitability. Cost-volume-profit analysis, which is discussed later, is useful in knowing how much quantity should be produced in earning certain profits.

An initial production plan is prepared and it is compared with available capacity. The steps regarding iterative procedure for production planning and capacity planning are as follows:

(*a*) A production plan is prepared initially considering demand forecasts and other factors.

(*b*) To meet the production plan, capacity requirements are evolved. Capacity may be in terms of machines, storage space, human resources, raw materials, capital required, etc.

(*c*) If available capacity is equal to required capacity, then production plan is satisfactory and may be used for exploring finer details at shop floor levels.

(*d*) If available capacity is not equal to required capacity, then two situations may arise:

I. Available Capacity > Required Capacity

In this case, if proposed initial production plan is followed, then there is certain idle capacity. As far as possible, idle capacity should be avoided and production plan may be revised so that the capacity is fully utilized.

II. Available Capacity < Required Capacity

In such a case, either the arrangement for additional capacity may be made or the production plan is adjusted as per the existing capacity. Depending on the prevailing situation, following measures are adopted to increase the existing capacity:

(*i*) Allow overtime to workers,

(*ii*) Some of the work such as fabrication of certain components may be given to outside vendors on sub-contract,

(*iii*) Procure additional equipment/machines to increase the capacity,

(*iv*) Source the additional capital to enhance the capacity.

By following iterative procedure, available capacity is made equivalent to required capacity. When production plant can be achieved by making use of the available capacity, then it is finalized for further details.

Planning at aggregate level is useful in generating master production schedule which will specifically state what kind of product is to be made, in how much quantity and in which periods. Master production schedule is further explored for material requirements planning at lower levels of product structure.

4.2. Production Planning and Control at Shop Floor

For effective utilization of the resources on the shop floor such as machines, storage space, human resources, raw materials, tools including jigs and fixtures, etc. production planning and control (PPC) becomes necessary. Techniques relevant to PPC are as follows:

(*a*) Routing: This refers to the route followed by the part to be manufactured on the shop floor. In other words, it is the sequence of machines through which the part will move different operations one after another.

For example, a part to be manufactured is shown below:

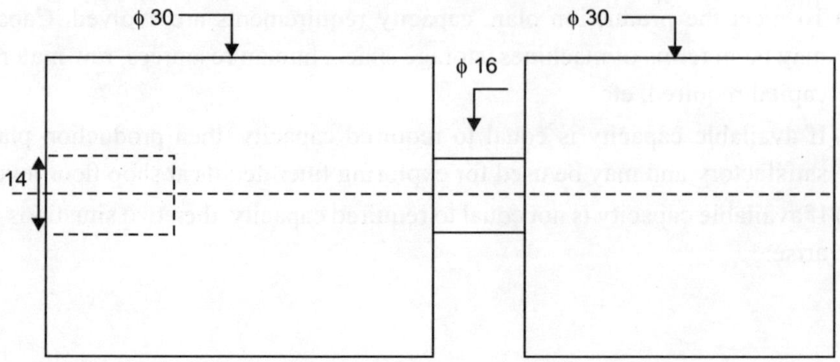

At the ends, the diameter of the part is 30 mm whereas in the middle, it has the diameter of 16 mm, and at one end it requires a drilled hole of diameter 14 mm. Application of the part also suggests grinding of the surface.

Following operations are needed to be performed:

(*i*) Turning

(*ii*) Drilling

(*iii*) Grinding

Sequence of operations suggests the part to move from lathe machine through drilling machine to grinder. The path to be moved by the product or part is decided and routing refers to that.

(***b***) **Scheduling:** Scheduling relates to time. It will state the time period in which specified quantity of the product/part is to be manufactured. Depending on the practice of the industry, schedule may be generated with respect to each machine/ worker. Each worker should be aware of the target, *i.e.* how much quantity is to be produced, in which time period (say day or week) and over which machine. The scheduling is also useful in knowing the load on machine or worker.

(***c***) **Despatching:** The routing and scheduling are basically planning tools whereas dispatching refers to execution of the planned work. As the planned work has to be converted into physical reality, all necessary orders for release of raw materials as well as tools including jigs and fixtures are issued. Job orders are issued so that all the resources are available in specified time period in order to complete the scheduled work successfully.

(***d***) **Follow-up:** To control the work, effective follow-up is essential. It is to be supervised that planned work is completed or not. If the work is in progress, whether it is being performed at the desired rate or not. A check is necessary for deviation in the target. Any deviation is analyzed thoroughly and remedial measures are implemented to ensure the completion of targeted production in time.

These four techniques may be applied for production planning and control. But on the shop floor, different performance measures need to be optimized with the help of certain rules or algorithm. For example in a job shop, if workers arrive at different times and there is a queue before a single machine. Which job is to be processed first and which job will be taken up later, *i.e.* sequencing of jobs on a machine is a problem to be solved.

4.3 Sequencing of Jobs on a Single Machine

It is also called as single machine scheduling. Various jobs are ready to be processed on a machine with individual due date of completion. Following are the parameters with respect to each job:

(*i*) **Ready time (R_j):** This is the time when job is ready for processing on a machines, in other words when it is released for processing.

(*ii*) **Due time (D_j):** It is the due time or due date for completion of any job$_j$. If the job is completed after due date, it gets delayed, *i.e.* it is *a* tardy job.

(*iii*) **Processing time (t_j):** It is the duration which a job will take on a machine, *i.e.* operation time.

From these input parameters, certain data are derived. The derived data are defined considering the following example:

Three jobs 1, 2 and 3 are to be processed on a machine in the same sequence. All the jobs are ready now, *i.e.* at time zero ($R_j = 0$). If due dates are provided, then the remaining time from now (*i.e.* at time zero) can easily by computed. Due time (D_j) is directly given as follows along with the processing time (t_j):

Job	D_j	t_j
1	2	1
2	2	1
3	5	2

(*i*) **Completion time (C_j):** This is the cumulative value, *i.e.* cumulative sum of t_j:

J	t_j	C_j
1	1	1
2	6	7
3	2	9

Completion time for the first job in the sequence is similar to processing time, *i.e.* $C_2 = C_1 + t_2 = 1 + 6 = 7$ and $C_3 = C_2 + t_3 = 7 + 2 = 9$. Therefore $C_j = C_j - 1 + t_j$.

Average completion, time, $\bar{F} = \dfrac{\sum Fj}{n}$ where n = number of jobs.

$$\bar{C} = \frac{1+7+9}{3} = 5.67 \cdot$$

(*ii*) **Flow time (F_j):** Flow time is the difference between completion time of a job, C_j, and ready time of that job, R_j.

$F_j = C_j - R_j$.

This is the time duration in which job flows on the shop floor. As R = 0, flow time is similar to completion time, *i.e.* $F_j = C_j$.

Average or mean flow time, $\bar{F} = \dfrac{\sum Fj}{n}$

Maximum flow time, $F_{max.} = 9$

(*iii*) Queue time (Q_j): This is the duration in which the job remains in queue before actual processing:

$Q_j = F_j - t_j$.

J	t_j	F_j	Q_j
1	1	1	0
2	6	7	1
3	2	9	7

Average queue time, $\bar{Q} = \dfrac{\sum Qj}{n} = \dfrac{0+1+7}{3} = 2.67$.

Maximum queue time, Q_{max} is 7 corresponding to job 3.

(*iv*) Number of Jobs in Process during certain time (P_n): This relates to work-in-process. Three jobs have been released to the shop floor before a machine M.

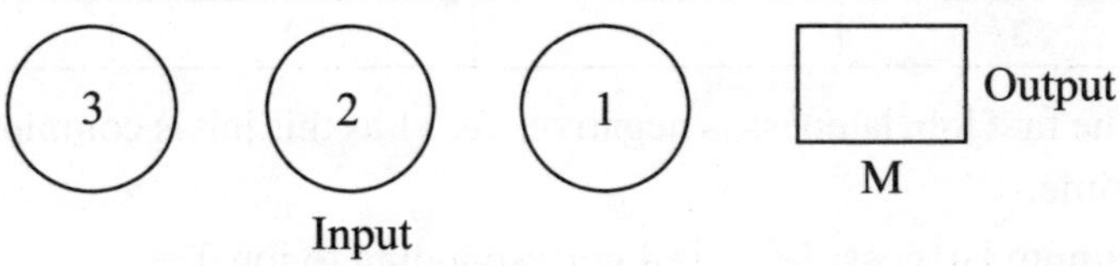

Now job (1) is processed on machine M for 1 unit of time as $t_i = 1$.

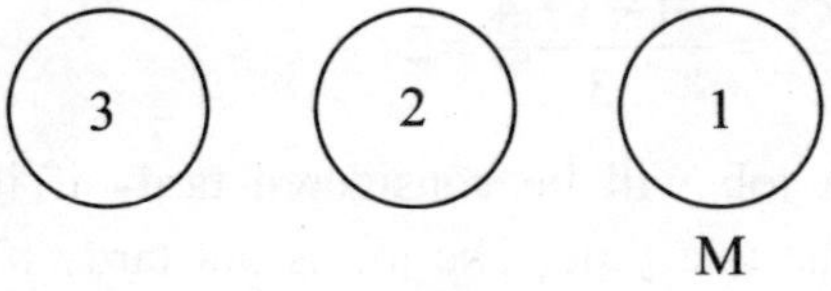

Three jobs are in process for time 1. Now processing of job 1 is over and job 2 will be processed on machine for 6 units of time.

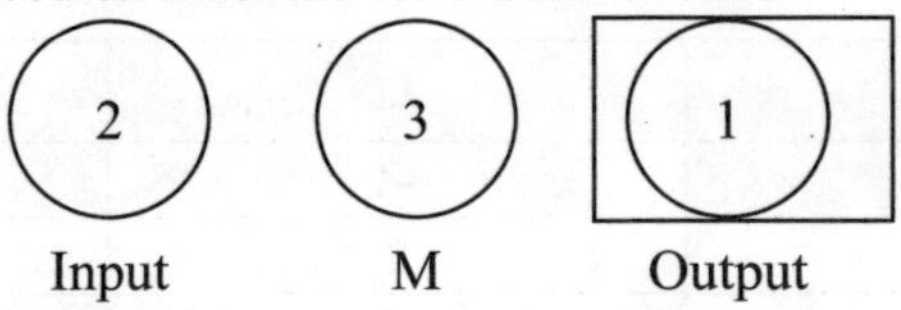

Two jobs (*viz*. 2 and 3) are in process for 6 units of time. Finally, one job, *i.e.* job 3 will be in process for 2 units of time, $t_3 = 2$.

Jobs are considered to be in process when either they await processing or are under actual operation over machine.

Now 3 jobs are in process for 1 unit of time as $t_1 = 1$.

2 jobs are in process for 2 units of time as $t_2 = 6$.

1 job is in process for 2 units of time as $t_3 = 2$.

Jobs will continue processing for maximum flow time, *i.e.* $F_{max} = 9$.

Therefore mean number of jobs in process

$$\overline{P} = \frac{(3\times1)+(2\times6)+(1\times2)}{9} = 1.89$$

or $$\overline{P} = \overline{p}\frac{\sum(\text{No. of jobs} \times \text{time for which they are in process})}{F_{max}}$$

If there are n jobs,

$$\overline{P} = \frac{[n \times t_1]+[(n-1) \times t_2]+[(n-2) \times t_3]+ \ldots\ldots + [2 \times t_{n-1}] + [1 \times t_n]}{F_{\max}}.$$

(*v*) Lateness (L_j): This is a measure by which it can be known whether a job is delayed or is it completed before the due time.

$L_j = C_j - D_j$.

J	D_j	C_j	$L_j = C_j - D_j$
1	2	1	–1
2	4	7	3
3	5	9	4

For the first job, lateness is negative, *i.e.* –1 as this job is completed before the due time.

Maximum lateness, L_{max}, is 4 corresponding to job 3.

Average lateness, $\overline{L} = \dfrac{-1+3+4}{3} = 2$.

(*vi*) Tardiness (T_j): A job will be considered tardy if it is delayed, *i.e.* it is completed after the due time. The job is not tardy if it is completed on or before the due date. If lateness is positive, then tardiness T_j is similar to L_j (otherwise T_j is zero).

J	L_j	T_j
1	–1	0
2	3	3
3	4	4

For L_j being negative or zero, tardiness, T_j is zero (otherwise $T_j = L_j$).

Considering the column T_j (values 0, 3, 4), maximum tardiness, T_{max}, is 4 corresponding to job 3.

Different measures of performance are as follows:

Average completion time – $\overline{P}$,

Average flow time – $\overline{F}$,

Average queue time – $\overline{Q}$,

Maximum queue time – Q_{max},

Mean number of jobs in process, P

Average lateness – $\overline{L}$,

Maximum lateness – L_{max},

Maximum tardiness – T_{max}.

4.3.1 SPT Rule

Shortest processing time (SPT) rule suggests that the job with the shortest processing time should be sequenced first. In other words, the jobs are sequenced for operations on a machine in the order of increasing processing time. Jobs are arranged in the increasing order of processing time t_j.

Following measures of performance are minimized using the SPT rule assuming that all the jobs are ready at the same time:

average completion time – $\overline{C}$,

average flow time – $\overline{F}$,

average queue time – $\overline{Q}$,

maximum queue time – Q_{max},

mean number of jobs in process – $\overline{P}$,

average lateness – $\overline{L}$.

Example 4.1

The following seven jobs are ready at the same time, *i.e.* at time zero for processing on a single machine. Apply the SPT rule for sequencing of jobs and evaluate different measures of performance which will be minimized:

J:	1	2	3	4	5	6	7
D_j:	13	9	4	23	6	19	8
t_j	8	3	2	11	4	10	6

Now to apply the SPT rule, jobs are arranged in the increasing order of processing time t_j:

J:	3	2	5	7	1	6	4
D_j:	4	9	6	8	13	19	23
t_j:	2	3	4	6	8	10	11
C_j:	2	5	9	15	23	33	44

F_j:	2	5	9	15	23	33	44
Q_j:	0	2	5	9	15	23	33
L_j:	-2	-4	3	7	10	14	21

Measures of performance which will be minimized are evaluated as follows:

(*a*) Average completion time, $\bar{C} = \dfrac{\sum Cj}{n} = \dfrac{131}{7} = 18.71$.

(*b*) Average flow time, $\bar{F} = \dfrac{\sum Fj}{n} = \dfrac{131}{7} = 18.71$.

(*c*) Average queue time, $\bar{Q} = \dfrac{\sum Qj}{n} = \dfrac{87}{7} = 12.43$.

(*d*) Maximum queue time, $Q_{max} = 33$

(*e*) As $n = 7$ and as the sequence of job is 3 – 2 – 5 – 7 – 1 – 6 – 4,

Mean number of jobs in process,

$$\bar{P} = \frac{(7xt_3) + (6xt_2) + (5xt_5) + (4xt_7) + (3xt_1) + (2xt_6) + (1xt_4)}{F_{\max}}$$

$$= \frac{(7x2) + (6x3) + (5x4) + (4x6) + (3x8) + (2x10) + (1x11)}{44}$$

$$= \frac{131}{44} = 2.98$$

(*f*) Average lateness, $\bar{L} = \dfrac{\sum Lj}{n}$

$$= \frac{49}{7} = 7$$

$$= \frac{-2 - 4 + 3 + 7 + 10 + 14 + 21}{7}.$$

4.3.2 EDD Order

Jobs are sequenced in the order of their earliest due date (EDD) first. Jobs are arranged in the increasing order of due date or due time, D_j.

EDD order minimizes the following measures of performance:

(*i*) Maximum lateness – L_{max}

(*ii*) Maximum tardiness – T_{max}

Example 4.2

Consider the input data of example, 4.1 and sequence the jobs in EDD order. Evaluate the performance measures which will be minimized using EDD.

Jobs are arranged in EDD order as follows:

J:	3	5	7	2	1	6	4
D_j:	4	6	8	9	13	19	23
T_j:	2	4	6	3	8	10	11
C_j:	2	6	12	15	23	33	44
L_j:	-2	0	4	6	10	14	21
T_j:	0	0	4	6	10	14	21

$L_{max} = 21$ and $T_{max} = 21$ whereas average lateness, $\overline{L}$, is obtained as 7.57.

4.4 Method to Minimize Number of Tardy Jobs

If the objective is to minimize the number of tardy jobs, then following method is adopted:

(1) Imagine two locations X and Y. X represents the set of jobs which are sequenced in EDD order and initially Y is an empty set of jobs.

(2) If no tardy jobs are found in X, then this is the desired sequence. Otherwise, the first tardy job in X is identified and this is called as the pth job.

(3) Now consider only the first P jobs in set X and the job with maximum processing time is picked among the first P jobs. This is shifted to another set Y. But in this process, calculations previously made will be disrupted. Completion time, C_j and tardiness, T_j, are recomputed.

The process from step (2) onwards is repeated till no tardy job in found in set X, *i.e.* all the jobs in set X will have tardiness T_j as zero.

Example 4.3

Consider the input data of example 4.1 and apply the method to minimize the number of tardy jobs.

As the arrangement of jobs in EDD order is the essential requirement to apply the iterative method, accordingly computations will be made. This process is already completed in example 4.2 and the computations are reproduced below. Sets X and Y are marked where Y is the empty set at present.

Table 4.1 Computation corresponding to EDD order

	X						
J:	3	5	7	2	1	6	4
D_j:	4	6	8	9	13	19	23
T_j:	2	4	6	3	8	10	11

C_j:	2	6	12	15	23	33	44
L_j:	-2	0	4	6	10	14	21
T_j:	0	0	4	6	10	14	21

Steps (2) and (3) are to be repeated in each cycle till all jobs in set X will have zero tardiness. Refer to Table 4.1. Only two jobs, *viz*. 3 & 5 are non-tardy jobs out of 7 jobs.

Cycle 1

Sequence of jobs 3 – 5 – 7 – 2 – 1 – 6 – 4 and the first tardy job is job 7 with $T_j = 4$; job 7 is third in the sequence,

$$\therefore p = 3.$$

Now consider only the first three jobs in the sequence, *i.e.* jobs 3, 5 & 7 with processing time t_j = 2, 4 and 6 respectively. Maximum processing time is 6 corresponding to job 7, therefore job 7 is shifted to set Y. Now the sequence is 3 – 5 – 2 – 1 – 6 – 4 – 7, recompute the parameters as shown in Table 4.2.

Table 4.2 Computation at the end of cycle 1

	X						*Y*
J:	3	5	2	1	6	4	7
D_j:	4	6	9	13	19	23	8
t_j:	2	4	3	8	10	11	6
C_j:	2	6	9	17	27	38	44
L_j:	-2	0	0	4	8	15	36
T_j:	0	0	0	4	8	15	36

Cycle 2

Consider Table 4.2 as the next iteration. First tardy job is 1 which is fourth in the sequence. Processing time of first four jobs are 2, 4, 3 & 8. Maximum processing time is 8 corresponding to job 1 and job 1 is shifted to set Y as shown in Table 4.3.

Table 4.3 Computation for cycle 2

	X					*Y*	
J:	3	5	2	6	4	7	1
D_j:	4	6	9	19	23	8	13
t_j:	2	4	3	10	11	6	8
C_j:	2	6	9	19	30	36	44
L_j:	-2	0	0	0	7	28	31
T_j:	0	0	0	0	7	28	31

Cycle 3

In table 4.3, the first tardy job is 4 which is 5th in the sequence. In the first five jobs, maximum processing time is 11 corresponding to job 4. Job 4 is shifted to Y as shown in Table 4.4.

Table 4.4 Computation for cycle 3

	X				Y		
J:	3	5	2	6	7	1	4
D_j:	4	6	9	19	8	13	23
T_j:	2	4	3	10	6	8	11
C_j:	2	6	9	19	25	33	44
L_j:	-2	0	0	0	17	20	21
T_j:	0	0	0	0	17	20	21

No tardy job is in set X, therefore this is the desired optimal sequence. The sequence is 3 – 5 – 2 – 6 – 7 – 1 – 4 with three tardy jobs 7, 1 and 4 with tardiness values, 17, 20 and 21 respectively.

4.5 Two-Machine Flow Shop Scheduling

As shown in Fig. 4.1, all the jobs are processed on machine 1 and machine 2 sequentially.

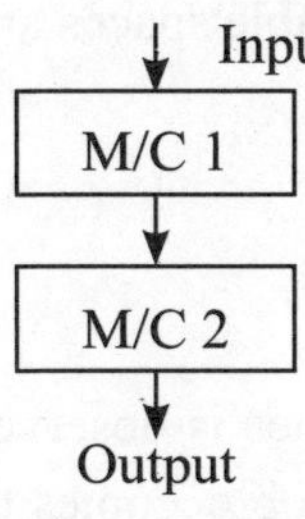

Fig 4.1 Two-machine pure flow shop

A number of jobs are on the input side of two-machine flow shop. Processing times of all jobs on both the machines are known. The time when the last job will be completed in the flow shop is of interest. This is called makespan, *i.e.* the maximum completion time, C_{max}. Objective is to determine the sequence of jobs in order to minimize the make span and Johnson's algorithm is used for this objective.

4.5.1 Johnson's Algorithm

Following steps are used:

(1) Processing times of all the jobs on each machine are observed and the least processing time is selected. If this least processing time corresponds to machine 1, go to step 2. If this least processing time is on machine 2, go to step 3.

(2) Job corresponding to the least processing time will occupy the earliest available space in the sequence. Go to step 4.

(3) Job corresponding to the least processing time will occupy the latest available space in the sequence.

(4) Keep the job, which has occupied the respective space in the sequence, out of consideration now. For the remaining jobs, repeat all the steps (1) through (4) until all jobs are accommodated in the sequence.

Example 4.4

A two machine flow shop is interested in sequencing of jobs so that makespan is minimized. Operation times of all six jobs on each machine are as follows:

Job	Time	
	Machine 1	Machine 2
J_1	8	5
J_2	6	7
J_3	4	8
J_4	15	14
J_5	8	9
J_6	11	10

As there are six jobs, six available spaces are there in the sequence as shown below:

Cycle 1

Operation time on both the machines is observed. Least time is 4 on machine 1 which corresponds to job j_3. This job occupies the earliest available space in the sequence: J_3 ------------------ .

As the job J_3 is placed in the sequence, this is eliminated.

Cycle 2

Remaining jobs are J_1, J_2, J_4, J_5 and J_6. From their processing times, least is 5 corresponding to job J_1. As this time is on machine 2, assign the job to the latest available space in the sequence: J_3--------J_1.

Cycle 3

Out of the processing time of the remaining jobs, J_2, J_4, J_5 and J_6 on both machines, 6 is the least time on machine 1 corresponding to job J_2 and it is placed at the earliest available space: J_3 J_2 J_3 ------ J_1

Cycle 4

Consider remaining jobs J_4, J_5 and J_6. Least time is 8 on machine 1 for job J_5, assign it to the earliest available space: $J_3\ J_2\ J_5$--- J_1.

Cycle 5

Remaining jobs are J_4 and J_6. Processing times are 15 & 14 on machines 1 & 2 respectively for J_4. 11 & 10 are processing time on machines 1 & 2 respectively for J_6. Out of processing time (15, 14, 11, 10), 10 is the least, corresponding to machine 2 for job J_6. J_6 is assigned to the latest available space: $J_3\ J_2\ J_5 - J_6\ J_1$.

Cycle 6

Only one job J_4 is left which will take the last available position in the sequence: $J_3\ J_2\ J_5\ J_4\ J_6\ J_1$

To minimize the makespan, the jobs are scheduled in the above order.

(**Note:** If there is a tie in selecting the least processing time, assign any one job arbitrarily)

Now, to evaluate the makespan, Gantt chart is prepared for this schedule as follows:

0 4 10 18 33 44 52

Machine 1	J_3 (4)	J_2 (4)	J_5 (8)	J_4 (15)	J_6 (11)	J_1 (8)

0 4 12 19 28 33 47 57 62

Machine 2	Idle	J_3 (8)	J_2 (7)	J_5 (9)	Idle	J_4 (14)	J_6 (10)	J_1 (5)

Individual time taken by each job on machine 1 and machine 2 are given in bracket along with jobs and cumulative values of time are shown on the top for each machine in the Gantt chart.

Last job in the sequence is J_1which will be completed at time 62. Therefore make span is 62 units of time.

Initially, machine 2 will be idle for 4 units of time as the first job, J_3 is being processed on machine 1. Machine 2 is also idle from time 28 to 33 as job, J_4 is not ready for processing. Job J_4 is completed on machine 1 at time 33 and then it will start processing on machine 2.

Sequencing of jobs is also summarized as follows:

Cycle	Remaining jobs to be scheduled	Least processing time	Sequence
1	J_1, J_2, J_3, J_4, J_5, J_6	4	J_3 ----------
2	J_1, J_2, J_4, J_5, J_6	5	J_3 -------J_1
3	J_2, J_4, J_5, J_6	6	$J_3\ J_2$ -----J_1
4	J_4, J_5, J_6	8	$J_3\ J_2\ J_5$ ---J_1
5	J_4, J_6	10	$J_3\ J_2\ J_5$—$J_6\ J_1$
6	J_4	14	$J_3\ J_2\ J_5\ J_4\ J_6\ J_1$

4.6 Cost–Volume–Profit Analysis

Quantity to be manufactured and sold is an important decision for production planning at aggregate higher level. Planned production should be in accordance with desired profit level. Cost–volume–profit analysis is useful for this purpose in addition to its other applications. Following are the types of costs which will be used for this analysis.

(*a*) **Fixed Cost:** It is concerned with the investments made in capital assets such as plant and machinery. This investment is fixed irrespective of the production quantity.

(*b*) **Variable Cost:** It is related to actual manufacturing and other associated costs. Total variable cost is proportional to quantity manufactured and sold.

(*c*) **Total Cost:** It is the sum of fixed and variable costs.

Fig 4.2 shows the graphical presentation of fixed, variable and total costs.

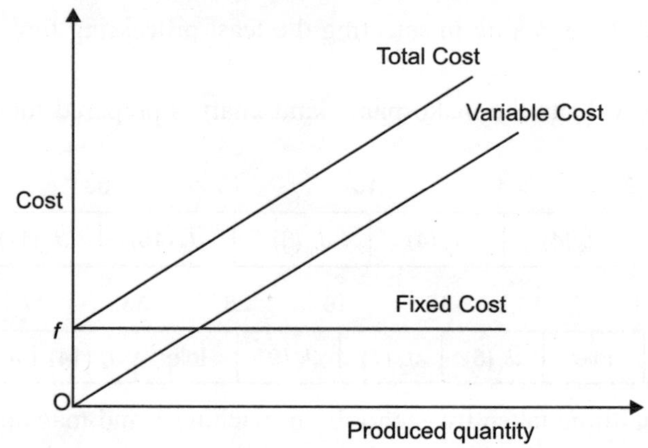

Fig. 4.2 Graphical presentation of costs

Let, fixed cost = f,

Variable cost per unit = v,

And quantity to be produced = q.

Fixed Cost, f, is represented by the horizontal line which is fixed irrespective of production quantity. Variable cost is shown by straight line passing through origin.

Variable cost = $v.q$,

And total cost = $f + vq$.

Total cost line is parallel to line for variable cost, passing through cost = f.

When production quantity is zero, total cost = f.

Now, if sales price per unit = s,

Then sales revenue = $s.q$.

Profit earned, p = sales revenue – total cost

Or $p = s.q - (f + vq)$,

Or $p + f = q(s - v)$

Or $q = \dfrac{f + p}{s - v}$

$(s - v)$ is called as contribution per unit.

If profit = o, then $q = \dfrac{f}{s - v}$ = Breakeven quantity

Break-even analysis is shown in Fig. 4.3.

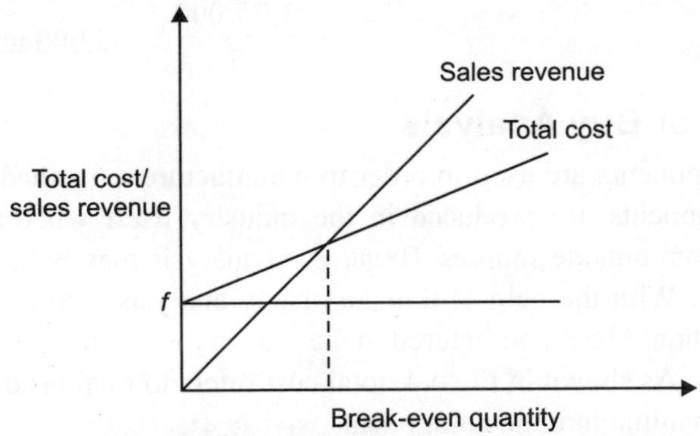

Fig. 4.3 Break-even analysis

If quantity to be manufactured and sold is less than the break-even quantity, then sales revenue is less than the total cost and there will be loss.

If quantity to be manufactured and sold is more than the break-even quantity, then sales revenue is greater than the total cost and there will be profit.

At the point of intersection of total costs and sales revenue line, *i.e.* at break-even point (BEP), there is no profit–no loss. At BEP, total cost is equal to sales revenue.

Example 4.5

The fixed cost of an industry is ₹ 2 lakh. Selling price per unit is ₹ 14 and variable cost to manufacture the item is ₹ 9 per unit. Obtain the BEP and contribution per unit.

Now f = ₹ 2×10^5,

s = ₹ 14,

And v = ₹ 9.

$$\text{BEP} = \frac{f}{s - v}$$

$$= \frac{2 \times 10^5}{(14 - 9)} = 40{,}000 \text{ units.}$$

BEP in terms of ₹ = 40,000 × 14 = ₹ 560,000.

Contribution per unit = $s - v$ = 14 – 9 = ₹ 5.

Example 4.6

Consider the data of example 4.5. If the industry wants to earn a profit of ₹ 75,000, how much quantity, should it produce and sale?

Now $p =$ ₹ 75,000

Quantity corresponding to profit, $p = \dfrac{f+p}{s-v}$

$$= \frac{2x10^5 + 75,000}{(14-9)}$$

$$= \frac{2,75,000}{5} = 55,000 \text{ units.}$$

4.7 Make or Buy Analysis

Several components are used in order to manufacture a finished assembly. Some of the components are produced in the industry itself whereas some may be purchased from outside sources. Break-even analysis may be used for a make or buy decision. With the help of a make or buy analysis, one can decide whether component should be manufactured in the industry or should it be purchased from other sources. As shown in Fig 4.4, total cost refers to total production cost in the industry to manufacture the component.

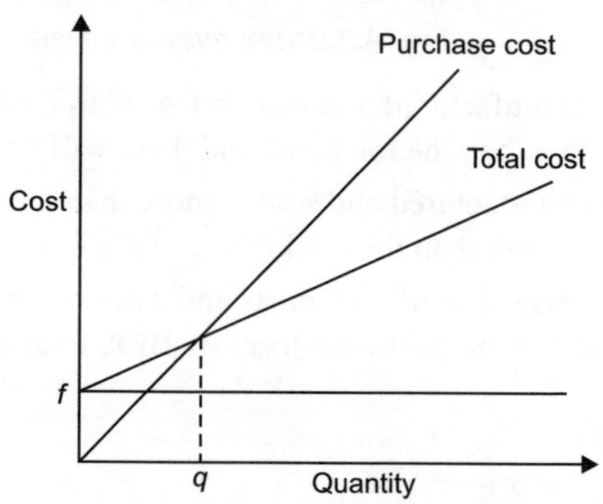

Fig. 4.4 Make or buy analysis

As discussed before, total production cost $= f + v.q$.

Where f is the fixed cost or investment corresponding to manufacture of the component and v is variable cost per unit.

If the component can be purchased from outside sources at the rate of s' per unit, then total purchasing cost $= s'.\ q$.

As shown in Fig. 4.4, q refers to the quantity corresponding to which total production cost is equivalent to total purchasing cost, *i.e.* $s'.\ q = f + vq$

or $q = \dfrac{f}{s'-v}$.

If the requirement of a component is less than q, purchasing cost is less than total manufacturing cost and therefore one should decide to buy the component. If quantity required is more than q, total production cost is less than the purchasing cost and therefore one should decide to manufacture the component in the industry.

Example 4.7

Fixed cost corresponding to manufacture of a component is estimated to be ₹ 50,000 by an industry. Variable cost is ₹ 12 per unit. At present the industry is purchasing the component at the rate of ₹ 22 each. Requirement of the component is 6,000 units. Decide whether component should be purchased or manufactured.

Now f = ₹ 50,000,

v = ₹ 12,

s' = ₹ 22

$$= \frac{f}{s'-v} = \frac{50,000}{(22-12)} = 5,000.$$

As the quantity required is 6,000 which is more than q, component should be manufactured in the industry itself.

Example 4.8

Show graphically how the concept of break-even analysis may be useful for taking a decision in replacing old equipment with new proposed equipment.

Let there be an existing equipment with fixed cost f_e and variable cost corresponding to it is v_e per unit. Total manufacturing cost with this equipment,

$TC_e = f_e + q.v_e$

Where q is the manufacturing quantity.

This existing old equipment is proposed to be replaced with new equipment.

f_n = Fixed cost corresponding to proposed new equipment.

v_n = variable cost per unit corresponding to new equipment.

Total manufacturing cost, $TC_n = f_n + q.v_n$

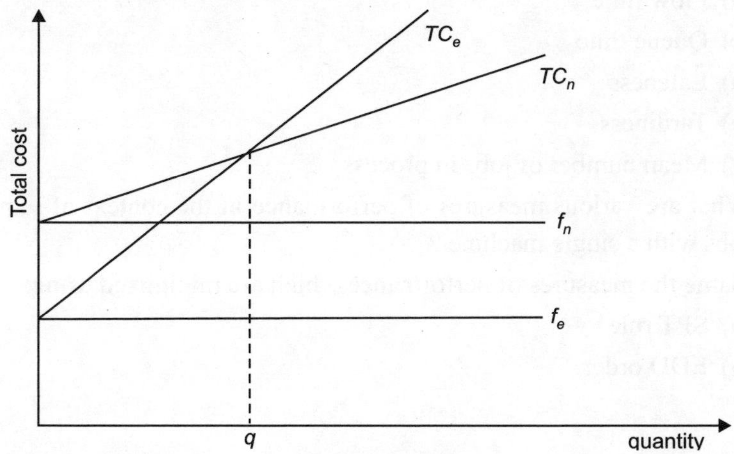

Fig. 4.5

Quantity q refers to the point of intersection of both the total cost lines *i.e.*, when:

$$TC_e = TC_{n'}$$

Or $$fe + q.v_e = f_n + q.v_{n'}$$

Or $$q = \frac{fn - fe}{ve - vn}.$$

If at present, the quantity to be manufactured is less than q, then $TC_e < TC_n$ and existing equipment need not to be replaced with new equipment.

EXERCISES

1. Differentiate between production planning and production control.
2. Explain the step regarding iterative procedure for production planning and capacity planning.
3. Discuss the following:
   (*a*) Routing
   (*b*) Scheduling
   (*c*) Dispatching
   (*d*) Follow-up
4. What do you understand by single-machine scheduling? Describe the input parameters with respect to any job.
5. Explain the following:
   (*a*) Completion time
   (*b*) Flow time
   (*c*) Queue time
   (*d*) Lateness
   (*e*) Tardiness
   (*f*) Mean number of jobs in process
6. What are various measures of performance in the context of sequencing of jobs with a single machine ?
7. Name the measures of performance which are minimized using:
   (*a*) SPT rule
   (*b*) EDD order

8. The following six jobs are ready at the same time, *i.e.* at time zero for processing on a single machine. Apply the shortest processing time (SPT) rule:

j:	1	2	3	4	5	6
D_j:	12	9	5	23	7	6
T_j:	6	3	4	13	2	5

Also evaluate measures of performance which will be minimized.

9. The following seven jobs are ready at time zero for operations on a single machine:

J:	1	2	3	4	5	6	7
D_j:	15	7	6	20	5	16	9
T_j:	9	2	3	15	4	9	8

Sequence the jobs in EDD order and evaluate the performance measures which will be minimized using EDD.

10. Explain the method to minimize the number of tardy jobs and apply it for the following data:

J:	1	2	3	4	5	6	7
D_j:	18	10	9	10	4	14	8
T_j:	13	6	7	17	2	10	6

11. (*a*) Write the steps involved in Johnson's algorithm.

(*b*) Consider a two-machine flow shop. Operation time of jobs on each machine are as follows:

Job	Time	
	M/C 1	M/C 2
J_1	7	6
J_2	8	9
J_3	3	9
J_4	13	11
J_5	5	10

Sequence the jobs so that makespan is minimized. Also construct the Gantt chart and comment on it.

12. Discuss the following in the context of break-even analysis:

(*a*) Fixed Cost

(*b*) Variable Cost

(*c*) Total Cost

Derive the break-even quantity and the quantity to be produced and sold to earn certain specified profit.

13. Consider the following data for an industry:

Fixed cost = ₹ 1.5 lakh,

Selling Price per unit = ₹ 19,

Variable cost per unit = ₹ 12.

Compute the approximate BEP and contribution per unit.

14. Explain how break-even analysis may be used for:

(*a*) Make or buy decision

(*b*) Equipment replacement decision

❑❑❑

5

Work Study

In general, productivity is defined as Output/Input. To increase the productivity, either output should be increased with the same level of input or input should be decreased for the same level of output. Productivity may also be increased by changing the output and input in such a way that the ratio of output to input is increased. For the profitability and improvement in the operations, continuous enhancement in the productivity becomes necessary. Work study is a set of techniques used to raise the productivity of an organization.

As the work study brings a change in the existing method of doing a task and normally human beings resist change, a suitable environment should be created before conducting the work study. Employees in the organization should be informed regarding the benefits of the work study, *i.e.* the productivity enhancement. By better utilization of resources, profitability will increase and with increased salaries and better growth prospects, it is for the welfare of employees. Work study is divided into two kinds of studies as follows:

Work study

Motion study (Method study) | Time study (Work measurement)

5.1 Motion Study

It is also called method study. Its objective is to develop more effective method and subsequent reduction in cost. This is achieved by systematic recording and critical examination of existing as well as proposed method of performing a task. Steps to be followed for conducting method study are discussed below:

(*a*) The whole job is to be divided into elements or smaller tasks,

(*b*) Each element of the industrial operation is to be examined critically. Some of the questions which may be asked are:

(*i*) Why this material is being used for processing? Is there any possibility of substituting this material?

(*ii*) Why is an operation conducted in a particular manner only? Is there any other way of doing this task?

(*iii*) Is there ease in operation if it is conducted on some other equipment or machine which is available?

(*c*) Critical examination of the operation will lead to modification in the existing method. Try to modify the existing method or develop a new method,

(*d*) Modified method or new proposed method needs training as well as arrangement of various facilities. Install the proposed method. Supervise the operations closely for a longer duration in order to maintain the proposed method. This is necessary because people usually revert to old practices.

5.1.1 Symbols used in motion study

The following five basic symbols are used in motion study:

Description	Symbol
Operation	O
Inspection	□
Delay	D
Storage	▽
Transport	⇨

Sometimes an operation is going on and simultaneously inspection may also be carried out. To represent this "operation and inspection", following symbol is used: 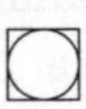.

5.2 Charts

There are certain charts which are useful in method study. These may be viewed as recording techniques. Following are various types of charts:

(*a*) Operation Chart: Operations and inspection to be carried out on any product are recorded as shown below:

Time Symbol Description

Suppose that a product requires four numbers of operations and inspection at two stages:

One, after operation 3 is over.

Another, after operation 4 is over which is also the final inspection.

After operation 3, the first inspection takes place that is why it is numbered as 1 in the symbol of inspection. On the left side, time taken for operation/inspection is recorded. On the right side, description of each operation/inspection is mentioned.

(*b*) **Flow Process Chart:** This chart indicates how product flows inside the industrial organization. But it may also be developed corresponding to activities of a person. Flow process chart is shown below:

Time taken	Distance travelled	Description	O	□	D	∇	⇨
—	—	Item in storage					
3 min	200 m	Item transported to shop floor					
15 min	—	Processing on the item					
4 min	—	Inspecting the item					
2 min	—	Item awaits for further movement					
3 min	200 m	Transporting the item to storage					
—	—	Item in store					

Initially an item is in storage, then it is transported to shop floor for processing. Transportation time is 3 min. and distance travelled is 200 m. Corresponding points below "storage" and "transport" symbols are joined by a straight line. Following the similar procedure, corresponding points are joined. Time taken in each activity as well as distance travelled, if any, are also recorded as shown along with description of each activity. If the item spends some time awaiting for any activity, it is recorded as 'delay'.

(*c*) **String Diagram:** Drawing of the layout using suitable scale is prepared on the drawing sheet. Drawing thus prepared is placed on the wooden board and pins which represent significant work centres are fixed at appropriate places on the sheet. Movement of the operator is recorded using a thread or string. Suppose that the person moves from work centre A to work centre B, then thread is tied to pin representing A and afterwards to the pin representing B on the drawing sheet. In this way, thread is tied as per the movement during specified period. The objective is to know how much distance is travelled between any two facilities, which may be determined by measuring the length of the thread considering the scale of the drawn layout.

String diagram is useful when more persons are moving in a certain area or a person has to go to two or more numbers of facilities often. If the total distance travelled between two facilities X and Y is very much and these two facilities are far apart, then they may be brought closer to each other if feasible.

(*d*) **Two-Hand Chart:** This is also known as left hand–right hand chart. The focus is on recording the functions of both the hands. The operation under consideration is studied by observing the movement of hands and time is recorded using a lock. Clock readings are recorded on the central portion of the paper and description of left–hand and right–hand motions are mentioned on left– and right–hand sides of the paper respectively.

(*e*) **Simo Chart:** Simultaneous motion (Simo) chart is similar to the two-hand chart except that the time needed for movement of the hands is recorded using a suitable time scale.

MICROMOTION STUDY

In the micromotion study, a permanent record in the form of a film is obtained using a camera with a clock. The film can be viewed any time later whenever needed and details of the operation can be observed.

CYCLEGRAPH

A small bulb is attached to any part of the body, say a finger, and film of the operation by worker is obtained with illuminated bulb. Movement of the finger or hand is highlighted which can be observed closely while viewing the film.

CHRONOCYCLEGRAPH

The bulb attached to the part of body is switched on and off frequently and the movement of the part is viewed as pear-shaped dots. By seeing the number of dots in a period or distance, one can decide whether the movement is fast or slow, or whether the part of the body is in acceleration or retardation.

(*f*) **Man–Machine Chart:** When an operator attends to one or more than one machine, then the interaction between man and machine is represented by the man–machine chart. It is also known as multiple activity chart. After constructing the man–machine chart, one can determine:

(*i*) For how much time, machine is idle,

(*ii*) For how much time, operator is idle.

Consider one operator and one machine. Loading time of machine 1 is 0.2 minutes and it needs to run for 0.3 minutes in order to complete the operation. Assume that unloading is instantaneous.

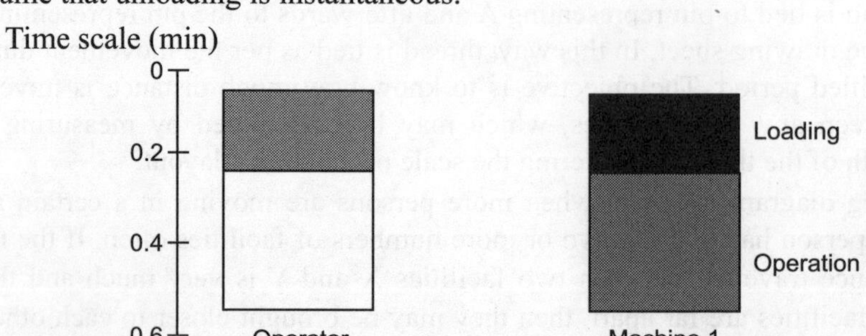

Fig 5.1 Man–machine chart for one operator and one machine

Time scale in minutes is shown in Fig. 5.1. Operator is engaged for 0.2 minutes in loading the machine. For the next 0.3 minutes, the operator is idle while machine is running. Cycle time is 0.5 minutes during which machine is either being loaded or in actual operation.

If a set of one operator and two machines is considered, then man–machine chart is drawn as shown in Figure 5.2.

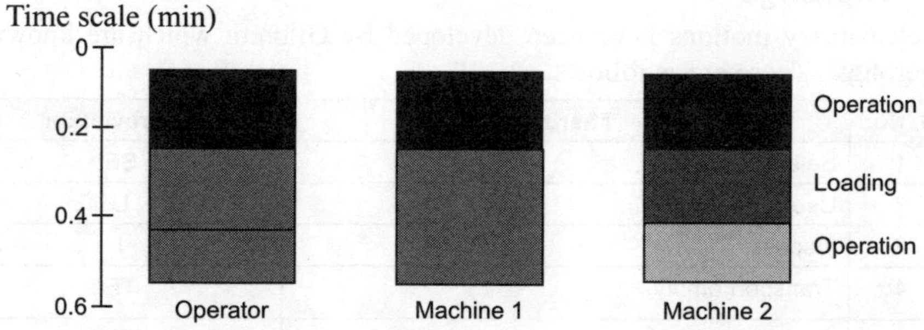

Fig. 5.2 Man–machine chart for one operator and two machines

Similar data are assumed for machine 2 also, *i.e.* loading time is 0.2 min. and operation time is 0.3 min. In a cycle time of 0.5 min., the idle time of operator is reduced to 0.1 min. For the first 0.2 min., the operator is loading machine 1, then for the next 0.2 min., he is loading machine 2, and idle for remaining 0.1 min.

Example 5.1

An item requires two successive operations which are planned to be done on two separate machines. Following data are given in minutes:

	Machine 1	Machine 2
Loading time	0.2	0.16
Operation time	0.2	0.21
Unloading time	0.1	0.1

Consider a set of one operator and two machines for the analysis and determine how many number of sets are required in order to produce 6,000 items per hour. Man–machine chart is drawn as follows:

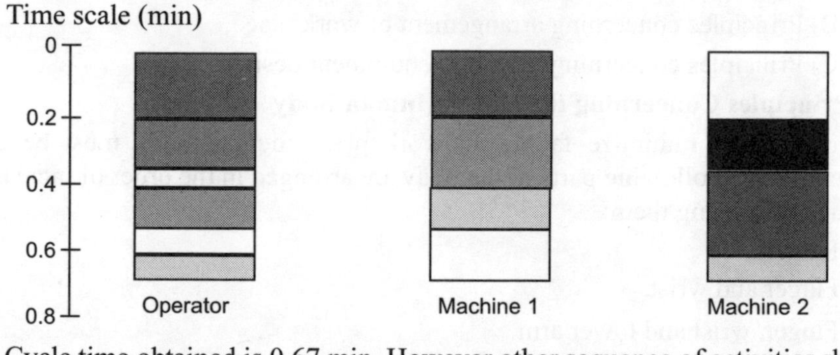

Cycle time obtained is 0.67 min. However other sequence of activities should be worked out which may yield a lower cycle time.

One set may produce 60/0.67 = 89.55 items per hour. Requirement is to produce 6,000 items per hour, therefore number of sets required = 6000/89.55 = 67.

In other words, 67 operators and 134 machines (67 of each type) are needed to produce the desired output.

5.3 Therbligs

17 elementary motions have been developed by Gilbreth which are known as 'Therbligs'. These are as follows:

S. No.	Therblig	Abbreviation
1	Select	SE
2	Use	U
3	Inspect	I
4	Transport empty	TE
5	Transport loaded	TL
6	Release load	RL
7	Pre-position	PP
8	Position	P
9	Disassemble	DA
10	Assemble	A
11	Plan	PN
12	Unavoidable delay	UD
13	Avoidable delay	AD
14	Rest to overcome fatigue	R
15	Search	S
16	Grasp	G
17	Hold	H

5.4 Principles of Motion Economy

All the principles of motion economy have been categorized mainly into three:

(A) Principles concerning the use of human body,

(B) Principles concerning arrangement of work place,

(C) Principles concerning tools and equipment design.

(A) Principles Concerning the Use of Human Body

In order to minimize fatigue at work place, human body must be used economically. Following parts of the body are arranged in the order of increasing fatigue while using them:

(*i*) Fingers,

(*ii*) Finger and wrist,

(*iii*) Finger, wrist and lower arm,

(*iv*) Finger, wrist, lower arm, upper arm and movement of whole body.

Rhythm needs to be generated at work, if possible, in order to enjoy the work and to decrease fatigue. Both the hands may simultaneously start and stop the movement, if possible. Movement should follow smooth curves instead of changing the direction abruptly.

(B) Principles Concerning Arrangement of Work Place

A pleasant environment should always be there at a work place. Temperature, humidity, illumination and ventilation should be at proper level. Eyes should not be strained while working including inspection of the components, etc.

Tools such as jigs/fixtures etc. should be arranged in such a way that searching and selecting them becomes easier and it takes as less time as possible. Clean surroundings always help in motion economy. Seating arrangement such as chairs for workers should be designed considering proper posture and comfort.

(C) Principles Concerning Tools and Equipment Design

Design of tools and equipment should incorporate issues related to human comfort. For this purpose, 'Ergonomics', *i.e.* human engineering is applied. For instance, handle of the tools such as screw driver, needs to be long enough (at least equivalent to the width of the palm). This will ensure comfortable use.

Similarly indicators of the equipment should be at the level of the eyes of the operator, as far as possible, in order to facilitate efficient viewing. Arrangement of illumination may be made at the place in the equipment where frequent inspection is needed. Provision of foot lever, if applicable, is useful from the point of view of motion economy.

5.5. Time Study

The objective of time study is to evaluate the standard time of completing a job or element of an operation. It is also known as work measurement.

5.5.1 Applications

Various applications of time study or work measurement are as follows:

(*a*) After knowing the standard time for any job, it is easier to estimate the cost involved in the job. Time spent on any job can be converted into money spent because people are paid on the basis of consumption of time in addition to other factors.

(*b*) Production planning becomes easier. With the help of standard time, production quantity in any period say day/week may be estimated.

(*c*) It also helps in production control. If the job is not finished, investigation is made and reasons are found for the deviation from the target.

(*d*) Productivity of the workers/employees is increased if wage/salary determination has a bearing on the time study.

5.5.2 Procedure to Conduct Time Study

Steps to be followed for time study are:

(*a*) Method of doing the job should be standardized before conducting time study. However there may be an iterative procedure. Based on the time study, the method is improved so that it takes less time, and then the improved method is standardized which may be used to conduct time study again.

(*b*) Entire task is divided into smaller elements after a thorough study. This exercise is useful for recording the observed time of each element.

(*c*) Average workers must be considered for conducting time study. If the selected workers are very fast, observed time and eventually standard time computed will be very less and most of the employees won't be able to complete the task in standard time. This causes dissatisfaction among the employees. On the other hand, if workers selected for the time study are very slow, standard time computed will be very high and productivity will be less. Therefore, workers selected should neither be very fast or very slow. These should be normal workers.

A few selected workers are made to do the same task and time taken is recorded with the help of a stopwatch. Average time is calculated which may be called as observed time.

Now, normal time (basic time) = observed time * performance rating factor.

Performance rating factor (PRF) or rating factor is used for leveling or normalizing the observed time. If an experienced industrial engineer feels that observed time is too high, then rating factor or PRF may be less than 1. Similarly, if observed time is too low, PRF may be more than 1.

Performance rating or PRF is the ratio of observed performance of the worker and normal performance from the industrial engineer's point of view.

Performance rating in % can be expressed as

$$\text{Performance rating (\%)} = \frac{\text{Observed performance of the worker} \times 100}{\text{Normal performance}}$$

After obtaining the normal time as discussed before, standard time is computed as follows:

Standard time (S.T.) = Normal time (N.T.) + Allowances,

Where allowances are expressed in terms of % of normal (or basic) time.

The allowance are of following types:

(*a*) Setup Allowance: Consider the following data:

Machine setup time = 60 minutes.

Once the machine is setup for a particular operation/process, it is ready for actual production. Normal time for production of one item on the machine is estimated as 7 minutes.

Due to the prevailing situations, industry is producing 60 items in one setup on that machine. Setup allowance for each item is the ratio of machine setup time and production volume in one setup which is 60/60 = 1 min.

This allowance of 1 min. can be expressed in terms of % of normal time as, 1/7 ×100 = 14.29%.

This is also called as special allowance which may also include other 'one-time' activities before the actual production starts. Depending on the circumstances the industry has to decide how to apportion this allowance on each item of production.

(*b*) **Allowance for Unoccupied Time:** If there are interruptions in the work due to reasons such as power failure, then allowance for unoccupied time should be included.

(*c*) **Interference Allowance:** If one operator is dealing with more than one machine, then one or more number of machines may wait for loading/unloading/starting. This idle time is accounted as interference allowance.

(*d*) **Relaxation Allowance:** Workers cannot be engaged in the work throughout the day/shift continuously because of personal needs such as drinking water. This allowance accounts for time spent in all such activities.

(*e*) **Contingency Allowance:** There are certain actions which are difficult to predict. For example, the worker has to discuss some technical issues with the superior for some time. Time, thus elapsed, may be included in contingency allowance.

Relevant allowances should be added to **arrive at the** value of total allowances.

Example 5.2

A job has been divided into four elements with following data:

Element	Observed time (min.)	Performance rating (%)
1	0.1	90
2	0.17	88
3	0.22	110
4	0.14	95

(*i*) Calculate the normal time for the job.

(*ii*) Compute the standard time for the job considering following allowances as % of normal time:

(*a*) Setup allowance 5%

(*b*) Relaxation allowance 10%

(*c*) Interference allowance 2%

(*i*) As the normal time = observed time × PRF,

PRF (performance rating factor) for element 1 = 0.90.

For element 2 = 0.88,

For element 3 = 1.10,

For element 4 = 0.95.

And normal time for element 1 = 0.1 * 0.90 = 0.09 min.

For element 2 = 0.17 * 0.88 = 0.1496 min,

For element 3 = 0.22 * 1.1 = 0.242 min,

For element 4 = 0.14 * 0.95 = 0.133 min.

Total normal time for the job = 0.6146 min.

(*ii*) Total allowances are 5 + 10 + 2 = 17% of the Normal Time.

Value of allowances = 0.6146 * 0.17 = 0.104482 min.

Standard Time = Normal Time + Allowances

= 0.6146 + 0.104482

= 0.719082 min

= 0.72 min.

5.5.3 Methods to Evaluate Performance Rating

As the performance rating is the ratio of observed performance and normal performance, speed of operator is judged with respect to normal pace of working. This is called 'speed rating'.

In addition to 'speed rating', following are additional methods to evaluate performance rating:

(*a*) Synthetic Rating: Predetermined motion time standards (PMTS) have been developed for various elements. These standards are considered as reference. PRF is evaluated as the ratio of predetermined standard for an element and observed time for that element as obtained in time study at work place.

(*b*) Westinghouse System of Rating: This is based on following four factors:

(*i*) Skill,

(*ii*) Effort,

(*iii*) Conditions,

(*iv*) Consistency.

Every job requires a certain level of skill in performing it. Level of skill is measured.

Efforts include physical as well as mental effort in performing a job. Degree of effort involved is decided.

Conditions refer to working conditions whether these are comfortable, hazardous, etc.

Consistency relates to level of performance with respect to time. It is observed whether performance (considering different parameters) is similar or deteriorates with respect to time.

In the literature, these factors have been graded such as 'excellent', 'good', 'average', 'fair', etc. and numbers are attached to each level.

Assume that for any task,

Skill corresponds to + 0.5,

Effort corresponds to + 0.06,

Conditions correspond to + 0.00,

And consistency correspond to + 0.02.

Total = + 0.13.

PRF = 1 + 0.13 = 1.13.

In other words, performance rating is 113%.

(*c*) **Objective Rating:** In addition to usual 'speed rating', allowance is also included for secondary adjustments such as job difficulties.

For example, with the use of 'speed rating', performance rating is 105%, *i.e.* 1.05%. Secondary adjustments are estimated as 125%, *i.e.* 1.25.

Finally, the PRF = 1.05 × 1.25 = 1.3125.

EXERCISES

1. What do you understand by productivity?

2. How is work study useful in increasing productivity?

3. Discuss the procedure for conducting method study.

4. What are the basic symbols used in motion study?

5. Explain the procedure of creating a flow process chart.

6. In which situation, is string diagram more useful?

7. What is micromotion study? Discuss the following:

(*a*) Cyclegraph,

(*b*) Chronocyclegraph.

❑❑❑

6

Inventory Management

In any industrial organization, materials are in the form of:

(*a*) Raw materials

(*b*) Work-in-process

(*c*) Finished products

These are called raw materials inventory, work-in-process inventory and finished products inventory. Significance of the materials management can be appreciated because material cost is almost 50–70% of the final product cost in most of the cases.

6.1 Benefits of the Inventory Management

Earlier the inventory control was considered to be a clerical task. But now-a-days, top management takes a lot of interest in the inventory control because it affects the probability of the concern in a significant way. It also helps in efficient running of the organization. Important benefits are as follows:

(*a*) It minimizes the total system cost which may include:

(*i*) Ordering cost

(*ii*) Inventory carrying cost

(*iii*) Purchase cost

(*iv*) Transportation cost

(*b*) Items are available as and when needed. In other words, right quantity is available at right time and right place.

(*c*) A decision regarding safety stock is taken so that suddenly increased demand is satisfied and shortages may be avoided if beneficial.

6.2 Economic Ordering Quantity

EOQ (economic ordering quantity) creates a trade-off between ordering cost and inventory carrying cost.

6.2.1 Ordering Cost

Suppose that an order for purchase of an item is placed with a supplier. Activities associated with placing an order may include:

(*i*) Spending some time in planning and writing the purchase order by an officer of the company

(*ii*) Using stationary, making telephone calls, etc.

(*iii*) Calling quotations and tender analysis

(*iv*) Selection of the supplier

Costs are involved in such activities and this is known as ordering cost. If larger quantities are ordered, then annual ordering cost is reduced. On the other hand, if less quantities are ordered, *i.e.* frequent ordering, then annual ordering will be more.

6.2.2 Inventory Carrying Cost

If certain quantities of an item are purchased, then inventory carrying cost may include:

(*i*) Storage cost

(*ii*) Obsolescence cost (if a large quantity is procured, then the item may became obsolete after some time and costs are incurred)

(*iii*) Investment is made in procuring the inventory, opportunity cost is the cost of lost opportunity (if investment is not made in inventories, it will be elsewhere which is a lost opportunity)

Inventory carrying cost increases with quantity procured. Fig. 6.1 shows the relationship of these types of costs with quantity, q, being ordered. Total cost is the sum of ordering and inventory carrying cost.

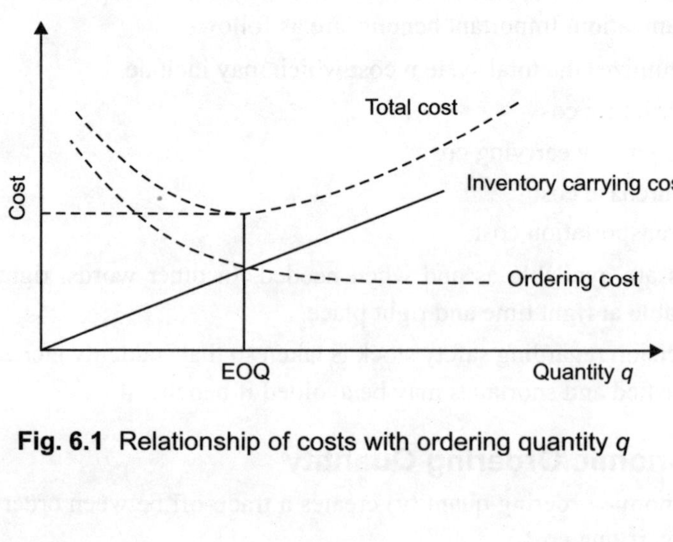

Fig. 6.1 Relationship of costs with ordering quantity q

Following consumption pattern is considered for EOQ formulation:

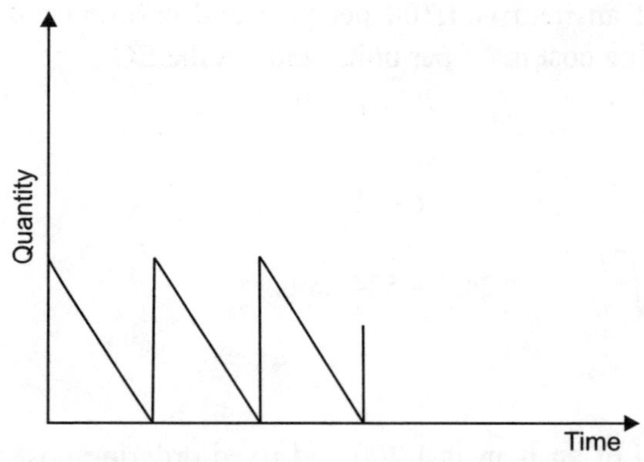

Fig. 6.2 Procurement of ordering quantity with respect to time

As shown in Fig. 6.2, quantity, q, is procured instantaneously and then it is consumed with demand rate, R, where R is the demand per year. As soon as the quantity in stock becomes zero another lot of quantity q is procured instantaneously and this cycle goes on with respect to time:

If C = ordering cost per order,

then annual ordering cost = $R/q.C$,

because number of orders placed in one year = R/q,

Quantity varies from q to zero. Average quantity is $q/2$.

If I = annual inventory carrying cost in ₹ per unit inventory, then annual inventory carrying cost $= q/2.I.$

Total annual cost, $E, = R/q.C + q/2.I$

EOQ corresponds to minimized total annual cost.

In order to obtain EOQ, differentiate E with respect to q partially and equating to zero:

$$\frac{\partial E}{\partial q} = 1,$$

or $$\frac{-RC}{q^2} + \frac{1}{2} = 0,$$

or $$\frac{RC}{q^2} = \frac{1}{2},$$

or $$\frac{RC}{q^2} = \frac{1}{2},$$

or $$q^* = \sqrt{\frac{2RC}{I}} = \text{EOQ}.$$

Example 6.1

Demand rate of an item is 1,100 per year and ordering cost is ₹ 250. Annual inventory carrying cost is ₹ I per unit. Find out the EOQ.

Now $R = 1{,}100,$

$C = 2.50,$

$I = 2.$

$$\textbf{EOQ} = \sqrt{\frac{2RC}{I}} = 524.4 = 524 \text{ units.}$$

Example 6.2

Annual demand of an item is 1,200 and fixed ordering cost per order is ₹ 250. Purchase cost of each item is ₹ 30. Considering annual inventory carrying cost fraction = 20%, determine EOQ.

Annual Inventory carrying cost = 20%.

Unit purchase cost = ₹ 30.

$\text{I} = 30 \times 0.2$

= ₹ 6 per unit-year.

$$\text{EOQ} = \sqrt{\frac{2 \times 1200 \times 250}{6}} = 316.23 = 316 \text{ units.}$$

Example 6.3

Show that annual ordering cost and inventory cost are equal corresponding to EOQ and derive the expression for minimum total cost:

Annual ordering cost = $R/q.C$

Substituting q as $\sqrt{\frac{2RC}{I}}$, *i.e.* EOQ,

Annual ordering cost $= \sqrt{\frac{RCI}{2}}.$

Annual inventory carrying cost = $q/2.I$

$= \sqrt{\frac{RCI}{2}}.$

Therefore annual ordering and carrying cost are equal corresponding to EOQ.

Minimum total annual cost $= \sqrt{\frac{RCI}{2}} + \sqrt{\frac{RCI}{2}} = \sqrt{2RCI}.$

6.3 Lead Time

An industrial organization places the purchase order with the supplier today and if the ordering quantity arrives after three days, then lead time is three days. Refer to Fig. 6.3. When level of quantity drops to a certain amount, then there is a need to place an order for that item. This level of quantity is called as reorder point (ROP).

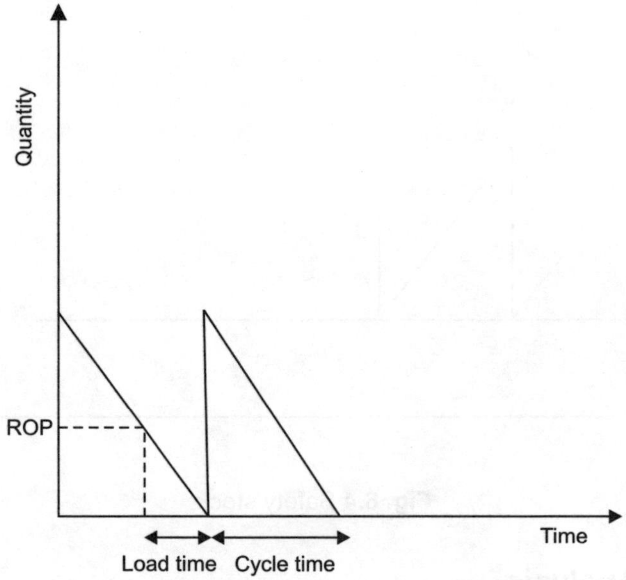

Fig. 6.3 Recorder Point and Lead Time

ROP = Lead Time × Demand Rate.

Example 6.4

If daily demand of an item is 4 and lead time is 3 days, find out the reorder point. Also determine the EOQ if cost of placing an order is ₹ 200 and daily inventory carrying cost is ₹ 1 per unit.

Calculate the cycle time.

As $$\text{ROP} = \text{L.T.} \times \text{demand rate},$$

$$\text{ROP} = 3 \times 4 = 12 \text{ units},$$

and $$\text{EOQ} = \sqrt{\frac{2 \times 4 \times 200}{1}} = 40 \text{ units}.$$

From Fig. 6.3, cycle time $= q/\text{demand rate} = 40/4$

$= 10$ days.

6.3.1 Safety Stock

It was assumed earlier that ordering quantity arrives as soon as the stock becomes zero. But in the real practice, a lot of quantity is not available when needed, then there will be shortage of item. As shown in Fig. 6.4, safety stock is needed to deal with uncertainties.

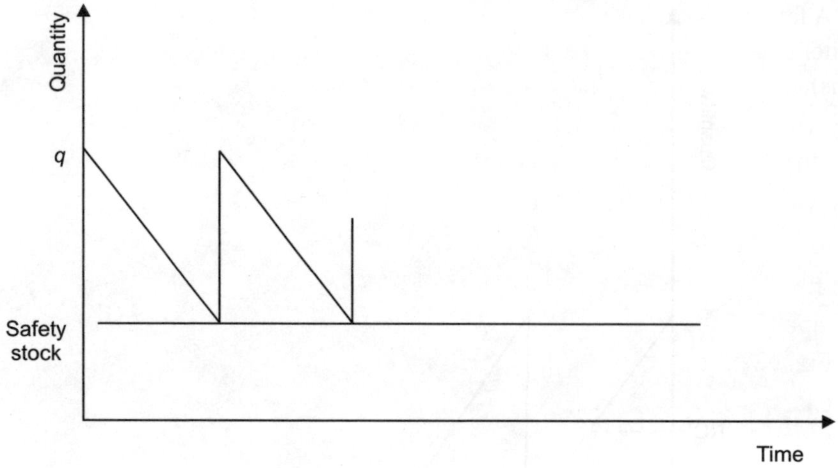

Fig. 6.4 Safety stock

6.4 ABC Analysis

It is difficult to pay equal attention to all the items in the industrial organization, therefore items are selected on the basis of certain variations and these are classified in various categories. ABC analysis is based on:

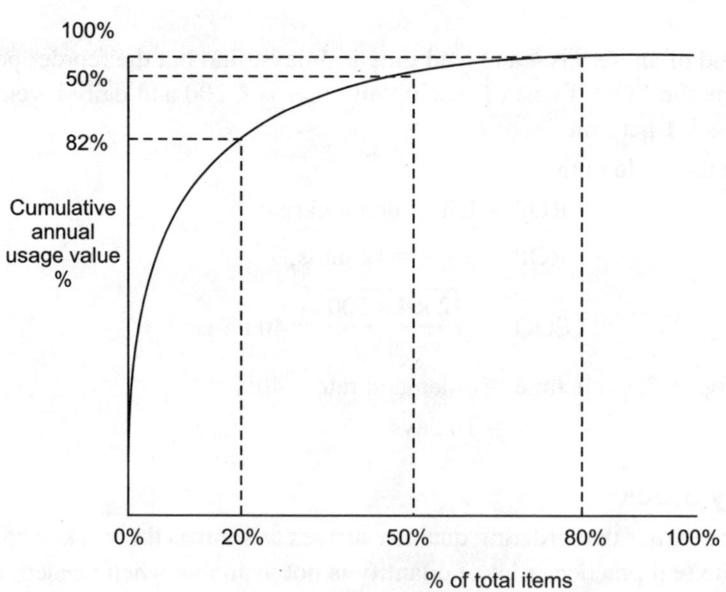

Fig. 6.5 ABC analysis

(*i*) Less number of items account for more value of annual usage

(*ii*) More number of items account for less value of annual usage

Annual usage value (or value of annual use) is the multiplication of annual consumption in units and price per unit item.

A few items of category 'A' refer to high annual usage value whereas category 'B' items account for medium value of annual use and large number (category 'C' items) account for low annual usage value.

(*a*) Approx. 10% of the total number of items ('A' items) account for approx. 70% of the total annual inventory value

(*b*) Approx. 20% of the total number of items ('B' items) relate to approx. 20% of the total annual inventory value

(*c*) Remaining approx. 70% of the total number of items ('C' items) refere to approx. 10% of the total value of annual usage.

Fig. 6.5 shows the ABC classification.

Classification in terms of % are indicative only. Depending on the data, certain variations may take place in the actual categorization of the items.

Example 6.5

An organization wants to analyze 15 items.

Annual consumption in units as well as price in ₹ per unit are given below.

Implement the ABC analysis for the given data and comment which of the items come under 'A', 'B' or 'C' category.

Item	Price/unit (₹)	Yearly consumption (units)
A	4	160
B	10	110
C	5	1030
D	4	65
E	5	130
F	20	45
G	10	4050
H	20	90
I	5	2000
J	50	14
K	3	800
L	2	300
M	3	3000
N	20	2005
O	30	40

Now annual usage value of each item is obtained by multiplying price per unit and annual consumption:

Item	Annual usage value
A	640
B	1100
C	5150

D	260
E	650
F	900
G	40500
H	1800
I	10000
J	700
K	2400
L	600
M	9000
N	40100
O	1200

Now arrange annual usage value in decreasing order, *i.e.* starting from the highest value to the lowest value, along with corresponding item as shown below:

Total annual inventory value = ₹ 1,15,000.

Now cumulative value of annual use is obtained such as 40,500 + 40,100 = 80600, 80600 + 10000 = 90600, ... This is also obtained in % of total inventory value such as (40500/115000) × 100 = 35.22%, (80600/115000) × 100 = 70.09%, ...

Category	Item	Annual usage value	Cumulative annual usage value	Cumulative value as % of total inventory value
A	G	40500	40500	35.22%
	N	40100	80600	70.09%
B	I	10000	90600	78.78%
	M	9000	99600	86.61%
	C	5150	104750	91.09%
C	K	2400	107150	93.17%
	H	1800	108950	94.74%
	O	1200	110150	95.78%
	B	1100	111250	96.74%
	F	900	112150	97.52%
	J	700	112850	98.13%
	E	650	113500	98.70%
	A	640	114140	99.25%
	L	600	114740	99.77%
	D	260	115000	100.00%

Out of total 15 items (*i.e.* from AV to O), two items G & N may be categorized as 'A' items because they constitute for approx. 70% of the total inventory value. 2/15 × 100 = 13.33% of the total items (namely G & N) are classified as 'A' category items which account for 70% of the total inventory value.

Three items, viz. I,M & C, *i.e.* $3/15 \times 100 = 20\%$ of the total number of items are categorized as 'B' items which account for 100 – 91.09 = approx. 9% of the total inventory value.

Remaining 10 items are 'C' items which are large in number, *i.e.* $10/15 \times 100 = 66.67\%$ of total items. These 'C' items account for 100-91.09 = approx. 9% of the total inventory value.

'A' items require utmost attention, *i.e.* they should be reviewed frequently for their stock status and decision regarding procurement. As these few items contribute for a major amount of inventory value, carrying cost is high and they may be purchased frequently or as and when needed. 'C' items which are large in number contribute for a small proportion of total inventory value, they may be purchased in bulk and frequency of reviewing them is very less. They require the least attention from managerial point of view. 'B' items are at immediate level. They require medium type of attention. Review period is greater than 'A' items and less than 'C' items.

In the real industrial environment, number of inventory items is very large. It is almost physically impossible to pay equal attention to all items individually. ABC analysis helps in categorization of inventory items and the management becomes convenient. In addition to ABC classification, items may also be analyzed as follows:

FSN Analysis

F stands for 'Fast'–moving items, S for 'Slow'–moving items and N for 'Non'–moving items. Suppose that an item of a particular design is obsolete and it has not been issued from the store for a long time. This can be identified by FSN analysis.

XYZ Analysis

This is based on high value, medium value and low value of an inventory item.

VED Analysis

Items are classified as 'Vital', 'Essential' and 'Desirable'. For example, from maintenance point of view, an item may be critical. For want of that spare part, the entire production process may be stopped. This may be classified as 'V', *i.e.* vital item and accordingly procurement strategy is decided.

SDE Analysis

SDE analysis is based on the degree of difficulty in procurement of an item.

S – Scarce item,

D – Difficult to procure,

E – Easily available.

The decision of how much to procure and when to procure may also be taken on the basis of scarcity of an item, easy availability may also be implemented.

6.5 Methods to Determine Value of Closing Stock/Pricing the Issues of Material

Purchase of the item/material and their issue for various purposes is a continuous process. Following two methods are discussed for pricing the material issue:

***(a)* FIFO Method:** First-in-first out (FIFO) method is explained with the following example:

(*i*) On September 9, 150 units of an item are procured @ ₹ 11 per unit

(*ii*) On September 12, 200 units of that item are procured @ ₹13 per unit

Now, on September 15, the item is to be issued for consumption and the quantity to be issued is 250.

On Sept. 9, value of the stock = 150 × 11 = ₹ 1,650

On Sept. 12, value of the stock = 1,650 + (200 × 13) = ₹ 4,250.

250 units of the item are to be issued on Sept. 15. As the method is first-in-first-out (FIFO), out of 250 units,

First 150 units are issued @ ₹ 11 per unit	=	1,650.
Remaining 100 are issued @ ₹ 13 per unit	=	1,300.
Value of the item issued	=	₹ 2,950.
Value of the stock on Sept. 15	=	4,250 – 2,950
	=	₹ 1,300.

***(b)* LIFO Method:** Last-in-first-out (LIFO) method can be applied for the example discussed before:

Out of 250 units which are to be issued on Sept. 15,

First 200 units are issued @ ₹ 13 per unit	=	2,600.
Remaining 50 units are issued @ ₹ 11 per unit	=	550.
Value of the item issued	=	₹ 3,150.
Value of the stock on Sept. 15	=	4,250 – 3,150
	=	₹ 1,100.

EXERCISES

1. What do you understand by 'Inventory'? Discuss the benefits of the inventory management.
2. What kind of trade-off takes place in determination of EOQ?
3. Discuss the assumptions involved in EOQ formulation and derive the EOQ.
4. Annual demand of an item is 1,300 and ordering cost is ₹ 200. If annual inventory carrying cost is ₹ 2.50 per unit, determine the EOQ.

5. Consider the following data:

Annual demand of the item = 1,250.

Fixed ordering cost = ₹ 300.

Unit purchase Cost = ₹ 25.

Annual inventory carrying cost fraction = 15%.

Determine the EOQ.

6. Derive the optimum annual total cost corresponding to EOQ. Show that annual inventory carrying cost and annual ordering cost are equal to each other with reference to economic ordering quantity.

7. Discuss the following:

(*a*) Lead time

(*b*) Reorder point, and their relationship

8. Determine the ROP if daily demand of an item is 5 and lead time is 2 days. If ordering cost = ₹ 400 and daily carrying cost of inventory = ₹ 0.4,

Calculate:

(*a*) EOQ

(*b*) Cycle time

9. (*a*) What is the benefit of having safety stock?

(*b*) Consumption of an item is 3 pieces per day. Fixed ordering cost is ₹ 50 per order. Annual inventory carrying cost is ₹ 1.50 per unit. Consider safety stock as 6 pieces and lead time as 4 days. Evaluate:

(*i*) Economic order quantity

(*ii*) Maximum stock

(*iii*) Reorder point

10. Discuss briefly the following:

(*a*) FSN Analysis

(*b*) VED Analysis

(*c*) SDE Analysis

11. Describe the ABC analysis in detail. Implement the ABC analysis for the following data:

Item	Price/unit (₹)	Yearly consumption (units)
A	3	200
B	11	100
C	5.5	1000
D	3.5	70
E	6	120
F	21	40
G	9	4100

H	19	100
I	4	2200
J	52	13
K	2	900
L	3	200
M	5	1600
N	19	2010

❑❑❑

7

Inspection and Quality Control

Quality of any item is related to fitness for its use. It is a kind of degree of perfection. Inspection is a procedure to measure the quality of an item which may be described by certain standards and specifications. Specifications are checked by inspecting the item and it is decided whether the item is accepted or rejected.

7.1 Inspection

Inspection is the detection of defect. Inspection department should be organized in such a manner that skilled staff is always available to discard the below–standard items. They need to be trained for using a variety of measuring instruments and in the upkeeping of the sophisticated equipment's. Different activities of the inspection division are as follows:

(*a*) As the material cost is more than 50% of the final product cost usually, it is important to inspect the raw materials thoroughly. If the raw material defects are detected at a later stage, entire cost of manufacturing may go waste.

(*b*) At every stage of manufacture, setting of the machine tool as well as first produced component should be inspected for defects and for different specifications. Once the first component is correct in all aspects, then further production should be allowed. Work-in-progress inspection is useful because problems are detected at that level only and defects are not passed to subsequent stages.

(*c*) The customers use the finished product marketed by the industry. Therefore finished products are checked in such a way that customers will be satisfied in using the product. In this way, customer complaints may be minimized. However, customer complaints cannot be eliminated completely and after receiving them, inspection division should be able to handle these efficiently.

7.1.1 Inspection on the Shop Floor *vs* Centralized Inspection

If the inspection process is performed on the shop floor itself, then inspector has to make a lot of movement. The inspector will go to shop floor at every facility and then check the components. Sophisticated equipment cannot be taken on the shop floor. But at the same time, decision making is fast.

In the centralized inspection, all the items/components are taken to a specified place in the factory for the inspection purpose. Inspection takes place in the suitable atmospheric conditions and modern equipment can be used. But the decision making is slow as items/components may queue up before centralized inspection facility.

A combination of these two ways of inspecting may be adopted to get the advantage of both the systems.

7.1.2 Sampling Inspection *vs* 100% Inspection

In case of 100% inspection, all the items are checked. It is a very monotonous task and due to fatigue, it may happen that a defective item sometimes passes through. However, wherever needed, 100% inspection should be adopted.

In sampling inspection, out of a lot, a sample of a few pieces is selected and the sample is checked for defects. If in the opinion of the inspector, the sample is good enough, then the entire lot is accepted. While selecting sample, care is to be taken that sample is a representative of the whole lot.

7.2 Acceptance Sampling

Suppose that in a lot of 1,000 items, one hundred items are picked up as sample, then sample size is 100. Depending on the inspection of sample of 100 items, fate of the lot of 1,000 items is decided. This is called acceptance sampling. There is a probability that a good lot is rejected and it is known as producer's risk. Similarly there is a probability that a bad lot is accepted which is called as consumer's risk. Fig. 7.1 shows operating characteristics (OC) curve.

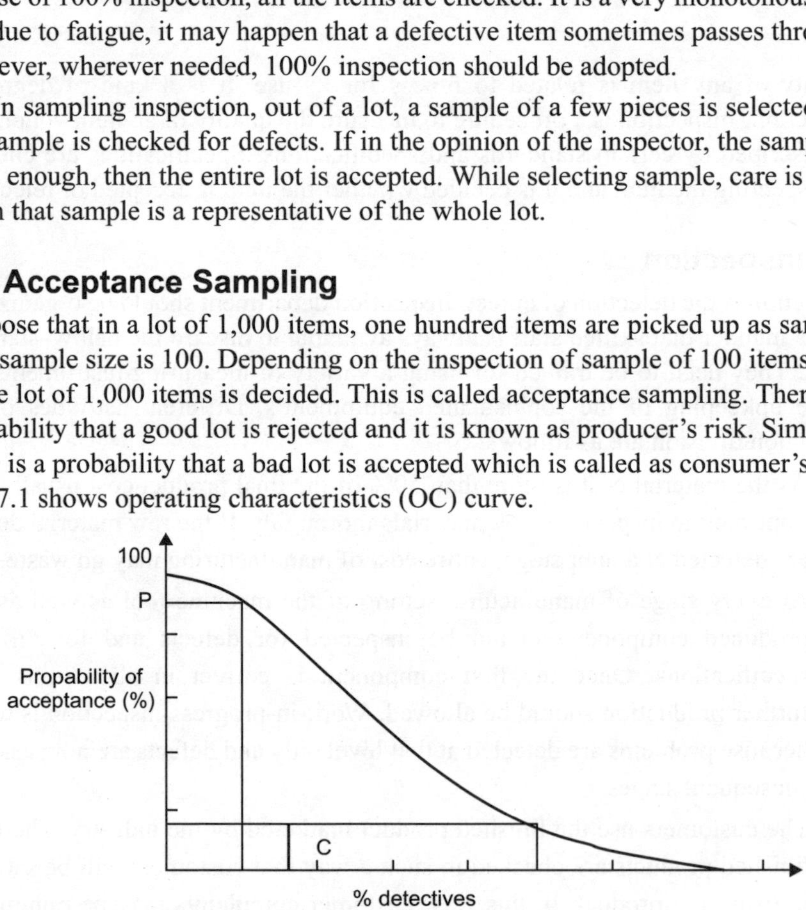

Fig. 7.1 OC curve

P = Producer's Risk.

This is the probability of rejecting a good lot (*i.e.* having less % defectives) on the basis of sample inspection.

C = Consumer's risk.

This is the probability of accepting a bad lot (*i.e.* having a large number of % defectives).

7.2.1 Single Sampling Plan

Assume that in a lot of 1,000 pieces, a sample of 100 pieces is picked up, *i.e.* sample size = 100. It is decided that after inspecting 100 pieces, if the number of defective pieces is less than or equal to 5, the whole lot of 1,000 pieces is accepted. In other words, acceptance number = 5.

To generalize, if sample size = S and acceptance number = A_n, then the lot is accepted if the number of defective pieces found in sample size S are less than or equal to A_n. The lot is rejected if the number of defective pieces are greater than An. In case of rejection, the lot is subjected to 100% inspection and all defective pieces found are either reworked or replaced with good pieces.

7.2.2 Double Sampling Plan

In the single sampling plan, a decision is taken after inspecting one sample only. But in double sampling plan, if need arises, second sample is inspected. Let first sample size = S_1 and second sample size = S_2. Acceptance corresponding to first sample = A_{n1}.

Acceptance number for both the samples = A_{n_2}.

First sample is inspected. Three situations that may arise are as follows:

(*i*) Number of defects <= A_{n_1}. In such a case, the lot is accepted.

(*ii*) Number of defects >A_{n_2}. The lot is rejected in this case.

(*iii*) In situations (*i*) and (*ii*) above, fate of the lot is decided and obviously there is no need for the second sample. But in third case, *i.e.*

Number of defects >A_{n_1}. But number of defects < = A_{n_2}. In such a case, the second sample is inspected.

If sum of the number of defects in the first and second samples <= A_{n_2}, the lot is acceptable. Otherwise if the total number of defects in both the samples >A_{n_2}, then the lot is finally rejected.

For example, A_{n_1} = 4 and A_{n_2} = 7.

First sample is inspected and if defects are up to 4, lot is accepted. If defective pieces exceed 7, lot is rejected. In case, number of defects is either 5, 6 or 7, then second sample is inspected. If combined number of defects > 7, lot is rejected. In case, this is up to 7, lot is accepted.

Similarly multiple or sequential sampling plan can be designed.

7.3 Quality Control

An item of diameter 15 mm is to be manufactured in large quantities. As one can imagine, it is difficult to manufacture every item with diameter 15 mm. exactly due to variation in the process and skill of the human resources. If tolerances are provided as 0.1 mm, then efforts are made to produce items with dimension ranging between 14.9 to 15.1 mm. However, overall objective should be to manufacture at nominal value (*i.e.* 15 mm) because deviation from nominal causes a loss.

Kinds of variation in a process are as follows:

(*a*) **Unusual Variation:** These are also called special cause of variation or assignable cause variation. For instance, it is identified that machine setting was not proper and due to that variation in the process has occurred. Assignable cause variations should be eliminated completely in order to obtain a stable process.

(*b*) **Usual Variation:** This is also called as chance variation or common cause variation. Such variations are present in the system itself and responsibility for these cannot lie on the employees of the industry. A stable process will have only usual variation. Process improvements efforts are made when process becomes stable.

Benefits of stable process are:

(*i*) Maximum productivity with minimization of costs

(*ii*) Easier to predict performance and quality levels and knowledge of process capability

(*iii*) Measurement of the effect of change to be made in the system,

(*iv*) If desired, it is easier to change the tolerances due to availability of supporting data.

7.3.1 Statistical Quality Control

SQC (Statistical Quality Control) makes use of statistics in controlling the quality characteristic.

If n number of item are inspected and a certain quality parameter x is measured for each item, then central tendency or mean is given by:

$$\bar{x} = \sum x/n$$

Dispersion in the data is measured by standard deviation,

$$\sigma = \frac{\sqrt{\sum (x-\bar{x})^2}}{\sqrt{n}}$$

Frequency distribution for normal curve is shown below:

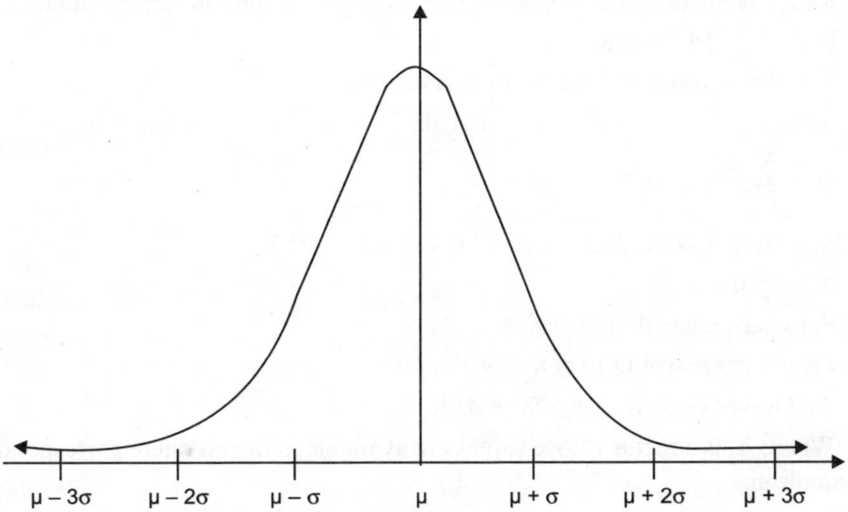

Fig. 7.2 Frequency distribution for normal curve

99.7% of the values fall within $\overline{x} \pm 3\sigma$,

95.5% of the values fall within $\overline{x} \pm 2\sigma$,

68.3% of the values fall within $\overline{x} \pm \sigma$.

7.4 Control Charts Variables

'Variables' refer to the situation when actual measurement such as 6.7 mm, 36 microns etc. are made for each item in the sample.

$\overline{X}$ and R–Charts: $\overline{x}$-chart and R-chart are useful for variables. Consider sample size = 4, *i.e.* in each sample, 4 items are included. Four measurements are recorded for each sample, say dimension in mm and seven samples are collected. Measurement, s, and calculations for $\overline{X}$ and R are shown below:

Sample No.	Measurement (mm)				Range	
	1	2	3	4	R	$\overline{X}$
1	15.5	14.7	15.2	14.9	0.8	15.07
2	15.1	14.8	14.9	15.3	0.5	15.02
3	14.6	14.9	15.4	14.8	0.8	14.92
4	15.2	14.8	14.9	14.9	0.4	14.95
5	15.3	15.2	15.3	14.8	0.5	15.15
6	14.7	14.9	15.4	14.8	0.7	14.95
7	15.1	15.1	14.8	14.9	0.3	14.98
					$\overline{R}$ = 0.57	$\overline{X}$ = 15.01

For example on sample No. 1, 4 items are measured and readings are 15.5, 14.7, 15.2 and 14.9 mm. Out of these four observations, maximum reading = 15.5.

Minimum reading = 14.7.

Range is the difference between maximum and minimum observation, *i.e.*

R = 15.5 –14.7 = 0.8.

$\overline{X}$ is the average of these four observations.

Similarly $\overline{X}$ and the R values are calculated as there are seven samples.

$$\overline{R} = \frac{\sum R}{7} = 0.57.$$

Similarly $\overline{X}$ is the average of all seven values of $\overline{X}$

$\overline{X}$ = 15.01.

For chart, control limit = $\overline{X} \pm A_2 \overline{R}$,

i.e. upper control limit = $\overline{X} + A_2 \overline{R}$,

And lower control limit = $\overline{X} \pm A_2 \overline{R}$.

Where A_2 is a factor whose values are available in the literature corresponding to sample size.

For sample size = 4, A_2 = 0.73

Upper control limit, UCL, = 15.01 + (0.73*0.57) = 15.42

Lower control limit, LCL, = 15.01 – (0.73*0.57) = 14.59

$\overline{X}$–chart is constructed as shown in Fig. 7.2.

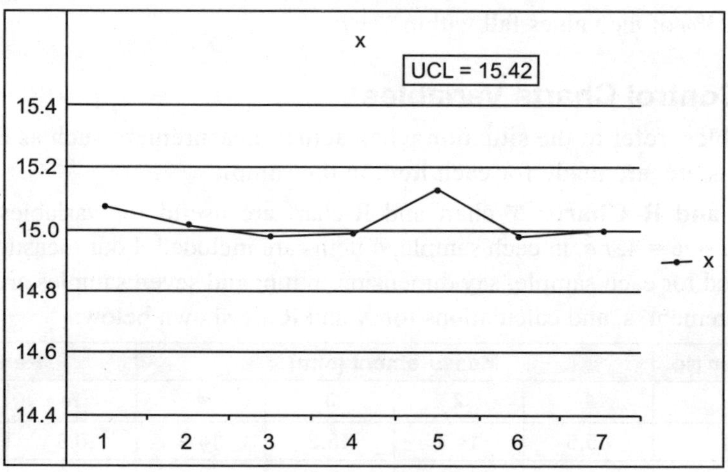

Fig. 7.2 $\overline{X}$-chart

For R–chart, UCL = $D_4\overline{R}$ and LCL = $D_3\overline{R}$ where D_3 and D_4 are factors whose values are available corresponding to sample size.

For sample size = 4, D_3 = 0, D_4 = 2.28.

UCL= 2.28*0.57 = 1.2996.

and LCL= 0*0.57 = 0.

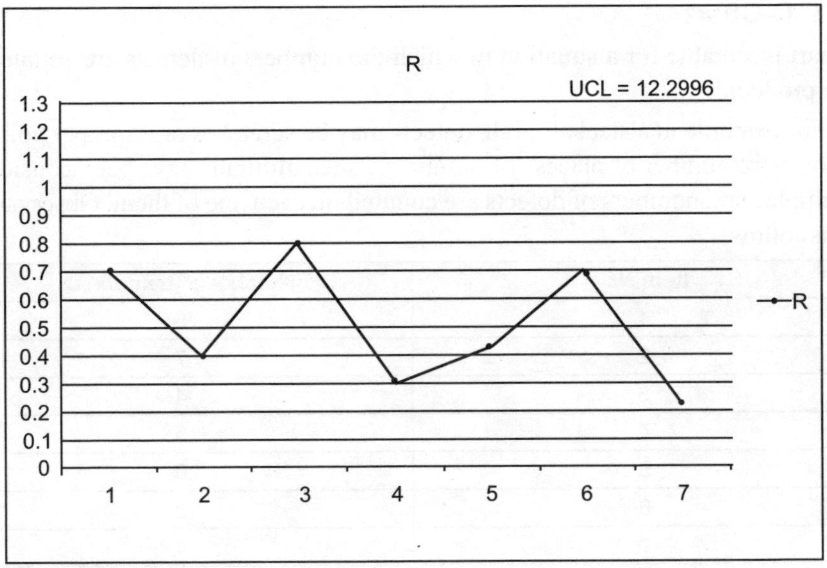

Fig. 7.3 R-chart

R-Chart is constructed as shown in Fig. 7.3. $\overline{X}$ and R values are well within the control limits. If for one or more samples, values fall outside the control limits, then further investigation becomes necessary. Reasons for the process going out of control occasionally should be identified and remedy is to be suggested for implementation.

After observing the chart, sometimes it may be concluded that $\overline{X}$ values are shifting towards one side that is either on UCL or LCL. It may be wear of machine parts or change in machine setting with respect to time. On the basis of this, frequency of correction in the machine to be determined and the process average which was tending to shift may be corrected to desired level.

7.5 Control Chart for Attributes

The attributes refers to a situation in which actual measurement is not taking place. The item is acceptable on the basis of certain criteria such as:

(1) Less than five number of scratches on the body of the automobile

(2) Unevenness of weld bead at less than or equal to on two points on the complete weld bead.

In addition to above, finding out the number of defective pieces in a sample using 'GO' and 'NO go' gauges also come under the category of attributes.

Total number of defects including scratches, dents, etc. on each product may also be recorded and analyzed.

7.5.1 C-Chart

C-chart is suitable for a situation in which the numbers of defects are counted on each product.

For example in a steel almirah, defects may be scratches or improper points at one or more number of places. 14 number of steel almirahs have been considered as samples and numbers of defects are counted on each one of them. Observations are as follows:

Item No.	Number of defects, C
1	6
2	7
3	4
4	5
5	3
6	2
7	4
8	6
9	5
10	6
11	3
12	1
13	2
14	2

Total number of defects = 56.

Mean number of defects $\bar{C} = 56/14 = 4$.

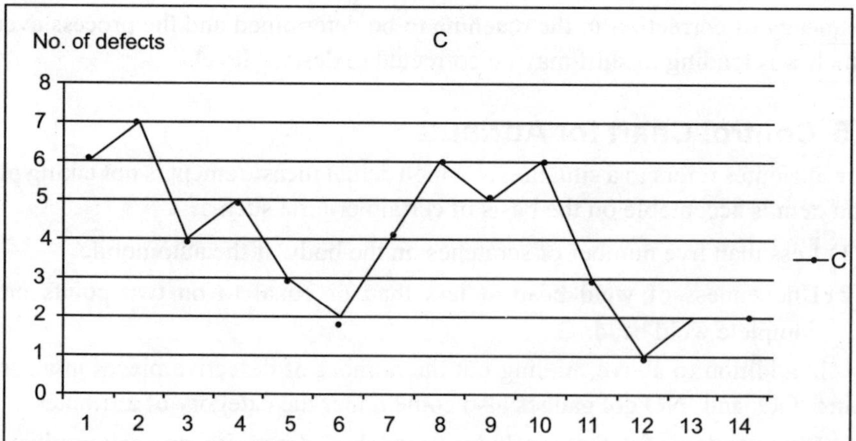

Fig. 7.4 C-chart

Control limit for C chart are:

$\bar{C} \pm \sqrt{\bar{C}}$,

i.e. $4 \pm 3\sqrt{4} = 4 \pm 6$

$UCL = 4 + 6 = 10$

$LCL = 4 - 6 = -2$

lower control limit is considered as zero because negative number of defects cannot exist. Now LCL = 0 and UCL = 10, C-chart is plotted as shown in Fig. 7.4.

7.5.2 P-Chart

P-chart is appropriate for a situation in which samples of some items are inspected and % defectives are obtained in each sample. Item is considered to be defective if it does not match the specifications.

For example, consider sample size =150.

Each sample includes 150 items. All 150 items in each sample are inspected and it is recorded that how many of them are defective items. Data collected for 12 samples and % defectives computation is given below:

Sample No.	Number of defectives out of 150 items inspected	% defective (P)
1	3	2
2	4	2.67
3	2	1.33
4	5	3.33
5	3	2
6	4	2.67
7	1	0.67
8	6	4
9	7	4.67
10	3	2
11	4	2.67
12	2	1.33

Total = 44 $\bar{P}$ =2.44.

For sample no. 1 % defective is obtained as 3*100/150 = 2%.

Similarly P values are obtained for each sample. Average P is obtained as $\sum P/12 = 2.44\% = \bar{P}$.

$\bar{P}$ can also be obtained as

$\bar{P}$ = total number of defectives/total number of items inspected,

Or $\bar{P} = 44/150*12 = 0.0244 = 2.44\%$.

Control limits for P-chart are:

$\bar{P} \pm 3\ \sigma_p.\ \bar{P}$

$\sigma_p = \sqrt{(\bar{P}\,(1 - \bar{P}\,))/N}$

N = Sample size = 150

$\sigma_p = \sqrt{(0.0244(1 - 0.0244))/150} = 0.0126$

Control limits are 0.0244 ± (3*0.0126).

UCL = 0.0622 = 6.22%

And LCL = – 0.0134, *i.e.* LCL = 0 as negative fraction defectives cannot exist.

P-chart may be constructed by plotting either fraction defective with respect to sample No. Fig. 7.5 shows the P-chart using % defectives.

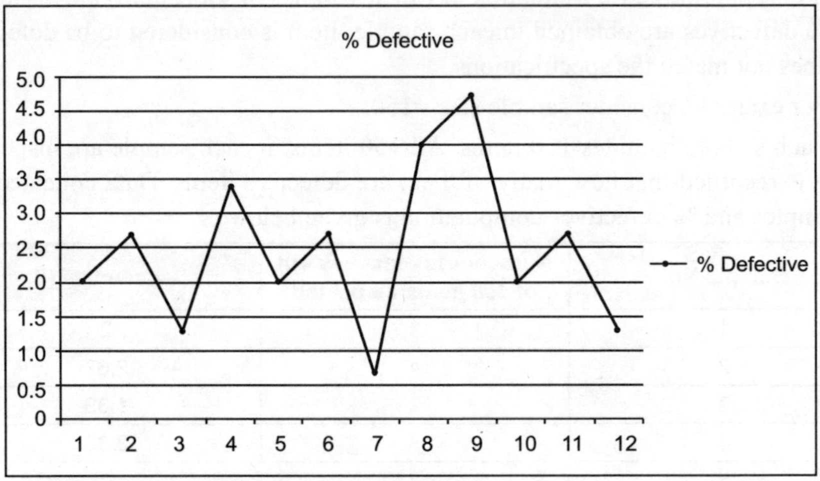

Fig. 7.6 P-chart

7.5.3 Six Sigma Criterion

UCL and LCL determined for the process are different from the tolerance specified for the product. If process verification can be reduced, it becomes more capable to produce acceptable products. Motorola adopted six sigma criterion by using tolerance specified for the product 6σ away from the process mean. This results into negligible defective products.

EXERCISES

1. What is inspection? Discuss different activities related to inspection.

2. Differentiate between:

(*a*) Centralized inspection and inspection on the shop floor

(*b*) Sampling inspection and 100% inspection

3. What is OC curve? Explain:

(*a*) Producer's risk

(*b*) Consumer's risk

4. Describe:

(*a*) Single sampling plan

(*b*) Double sampling plan

5. What are various kinds of variation in process?

6. Discuss the benefits of stable process.

7. What is S.Q.C.?

8. (*a*) Explain how to construct $\bar{X}$ and R charts,

(*b*) For which type of situations P are and C-charts useful?

9. Compute the control limits and construct the $\bar{X}$- and R-chart for the following data related to 10 samples with sample size = 5:

Sample No.	Measurement (mm)				
	1	2	3	4	5
1	7.7	7.3	7.6	7.4	7.9
2	8.5	8.0	7.9	7.6	8.2
3	8.1	7.8	8.3	7.7	8.0
4	7.9	8.0	7.8	8.3	8.2
5	8.3	8.1	7.7	7.8	8.1
6	8.2	8.2	7.9	8.0	7.8
7	7.8	8.1	7.7	7.5	7.8
8	8.4	8.3	7.6	7.5	7.5
9	7.7	8.1	8.3	7.9	7.8
10	8.0	8.3	8.4	7.6	7.9

Consider the following factors:

$A_2 = 0.58.$

$D_4 = 2.11.$

$D_3 = 0.$

10. Number of defects has been recorded for the following 12 items.

Item No.	Number of defects
1	8
2	9
3	7
4	8
5	6
6	5
7	8
8	6
9	4

10	3
11	2
12	7

Construct C-chart.

11. Ten samples have been inspected with sample size = 200. Construct the P-chart for the given data:

Sample No.	Number of defect items out of 200 items inspected
1	21
2	18
3	14
4	17
5	8
6	9
7	10
8	16
9	20
10	13

❑❑❑

8

Project Management

Construction of bridge, installation of plant and machinery, etc. refer to projects. A project will consist of several activities. Some activities may only run in parallel while some activities may only start when certain specific operations are over. Utilization of resources becomes an important issue so that the project is completed in time with minimum cost. Bar chart is a simple technique for project planning in which various activities are shown with respect to time (Fig. 8.1).

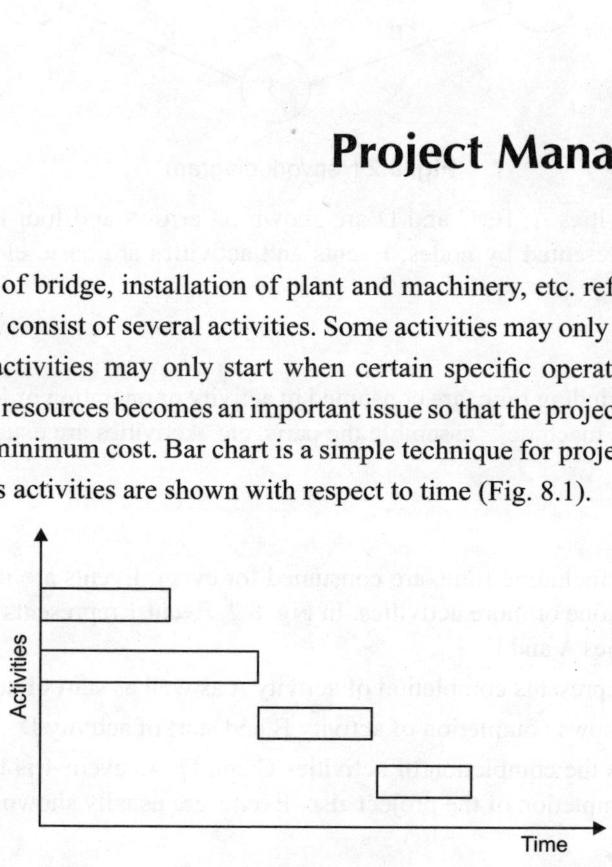

Fig. 8.1 Bar chart

Activities A, B, C and D are shown in Fig. 8.1. It can be determined when will an activity start and when will it end. But it is difficult to determine the interdependence among activities particularly when these are large in number.

Network methods are more useful techniques of project management as they are convenient to show interdependence among activities by drawing a network diagram.

A network diagram is shown in Fig. 8.2.

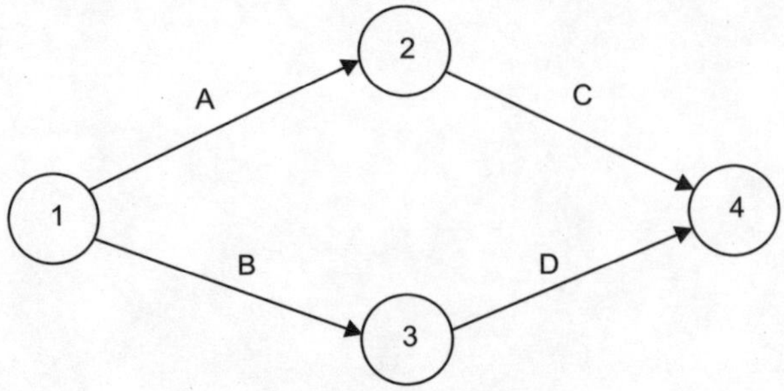

Fig. 8.2 Network diagram

Four activities A, B, C and D are shown on arrows and four events 1, 2, 3 and 4 are represented by nodes. Events and activities are basic elements of the network.

Activity

Resources, including time, are consumed in activity or operation or job. Examples are 'install the machine', 'assemble the parts, etc. Activities are usually shown on arrows.

Event

No resources, including time, are consumed for event. Events are just starting or completion of one or more activities. In Fig. 8.2, Event 1 represents simultaneous start of activities A and B.

Event 2 represents completion of activity A as well as start of activity C.

Event 3 shows completion of activity B and start of activity D.

Event 4 is the completion of activities C and D. As event 4 is the end event, this shows completion of the project also. Events are usually shown on nodes.

Example 8.1

Draw the network diagram for a small project with activities A, B, C, D, E, F and G. Activity-predecessor relationships are as follows:

Activity	Immediate predecessor
A	—
B	A
C	A
D	A
E	B
F	C
G	D

Completion of activities E, F and G can be treated as completion of the project.

Network diagram is drawn as shown below:

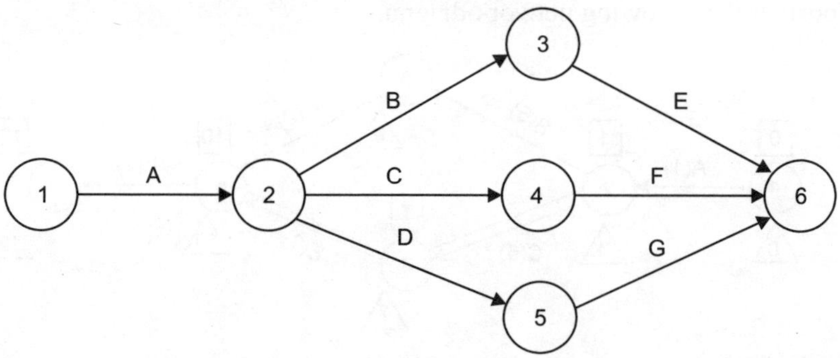

Event 1 is the starting event. As activity A is having no immediate predecessor, it is the first activity. Event 2 marks the completion of activity A. As A is immediate predecessor to three activities B, C & D, these activities emanate from Event 2 and their ircompletion marked by Events 3, 4 & 5 respectively. E, F & G follow activities B, C and D respectively. Activities E, F and G converge to end Event 6 because their completion shows the end of the project.

Dummy Activity

In example 8.1, if an additional condition is 'activity G cannot start until C is complete', then network diagram is drawn as shown in Fig. 8.3.

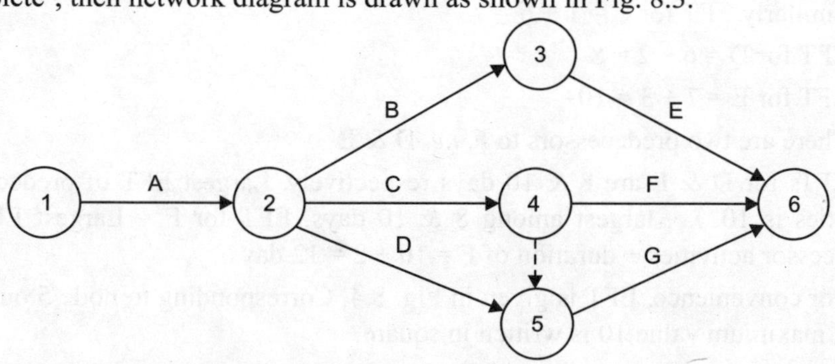

Fig. 8.3 Network diagram with dummy activity

Completion of C is Event 4 and start of G is Event 5. Events 4 and 5 are joined by the dotted line. 4–5 is the dummy activity which satisfies the condition that activity G cannot start until C is complete or Event 4 is over. Dummy activity does not consume resources, including time. It is just to show dependence among activities.

8.1 Critical Path Method (CPM)

CPM is activity-oriented technique of project management. It is suitable for the projects where duration of the activities can be estimated precisely and there is certainty in the activity duration.

Therefore, it is appropriate for deterministic situations and there is only one time estimate for each activity.

Consider the following network diagram:

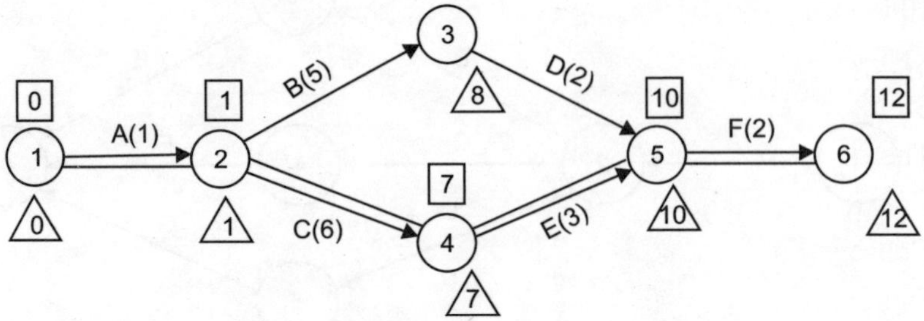

Fig. 8.4 Activity duration (days in bracket for activities A, B, C, D, E & F)

Earliest finish time (EFT) corresponding to starting event 1 is zero.

EFT for any activity = Largest EFT of predecessor activities + activity duration.

EFT for activity A = 0 + 1 = 1, as there are no predecessors to A and the activity duration of A is 1 day.

Since predecessor of B is A only, EFT for activity B = EFT of A + duration of B = 1 + 5 = 6.

Similarly EFT for C = 1 + 6 = 7.

EFT for D = 6 + 2 = 8.

EFT for E = 7 + 3 = 10.

There are two predecessors to F, *i.e.* D & E.

EFTs for D & E are 8 & 10 days respectively. Largest EFT of predecessor activities is 10, *i.e.* largest among 8 & 10 days. EFT for F = Largest EFT of predecessor activities + duration of F = 10 + 2 = 12 days.

For convenience, EFT is given in Fig. 8.4. Corresponding to node 5 out of 8 & 10, maximum value 10 is written in square.

To calculate EFT for each activity, forward movement was made with reference to Fig. 8.4.

Latest finish time (LFT) corresponding to last event = EFT corresponding to that event. 12 is written in Δ to specify LFT corresponding to event 6.

LFT for any activity = LFT of successor activity – duration of successor activity.

If there are more than one successor to any, activity, then the smallest value of LFT thus calculated will be chosen as LFT for that activity.

As F is the last activity, LFT for F = EFT = 12 days.

To calculate LFT for all activities, backward movement will be made corresponding to Fig. 8.4.

LFT for D = LFT of successor, *i.e.* F-duration of F = 12 – 2 = 10.

As the successor to E is also F,

LFT for E = 12 – 2 = 10.

10 is mentioned in A at node 5 to represent LFT for convenience in Fig. 8.4.

LFT for B = LFT of D-duration of D = 10 – 2 = 8

Similarly LFT for C = LFT of E-duration of E = 10 – 3 = 7

There are two successors to A, viz. B & C

Considering B, LFT for A = 8 – 5 = 3

Considering C, LFT for A = 7 – 6 = 1

Out of 3 & 1, smallest value, *i.e.* I will be chosen as LFT for A finally which is written in Fig. 8.4 corresponding to node 2.

For Event 1 which is the starting event, obviously 0 can be mentioned.

Total float = LFT – EFT.

Activities for which total float is zero, *i.e.* LETT = EFT are called critical activities and path having these critical activities is known as critical path.

EFT and LFT as well as T.F. (total float) or each activity are given as follows:

Activity	EFT	LFT	TF
A	1	1	0
B	6	8	2
C	7	7	0
D	8	10	2
E	10	10	0
F	12	12	0

Critical activities are A, C, E & F, therefore critical path is A—C—E—F shown as double arrows in Fig. 8.4. Total project duration = 12 days.

Earliest start time (EST) for any activity = EFT for that activity – activity duration.

Latest start time (LST) = LFT-activity duration. For example, duration of activity D = 2, EST for D = EFT for D-duration of D = 8 – 2 = 6 days and LST for D = LFT for D-duration of D = 10 – 2 = 8 days.

Similarly EST & LST are obtained for all the activities which are given below:

Activity	Activity duration	EST	LST
A	1	0	0
B	5	1	3
C	6	1	1
D	2	6	8
E	3	7	7
F	2	10	10

Total float can also be defined as (LST – EST).

Example 8.2

Following network diagram shows activities on nodes alongwith their duration:

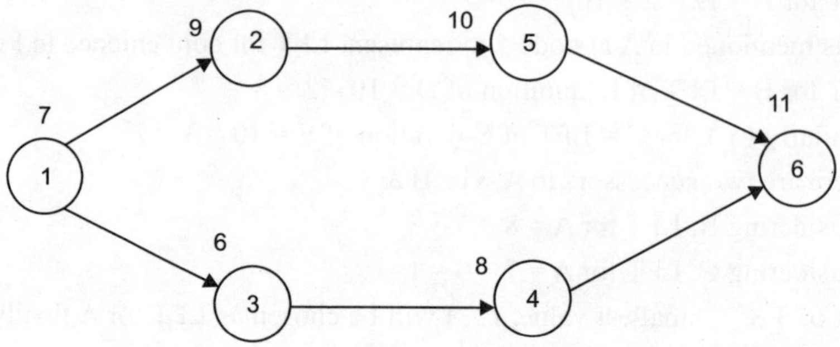

Determine the critical path and discuss the significance of critical path.

Activities 1, 2, 3, 4, 5 & 6 have been shown on the nodes. Calculations for EFT of all the activities are made while moving forward as discussed before. LFT is computed while moving backwards. The nodes on which EFT and LFT are equal (*i.e.* T.F. = zero) are joined by double lines to mark the critical path. The computation is summarized as represented below:

Activities 1, 2, 5 & 6 are critical activities and critical path is with overall completion time of the project as 37.

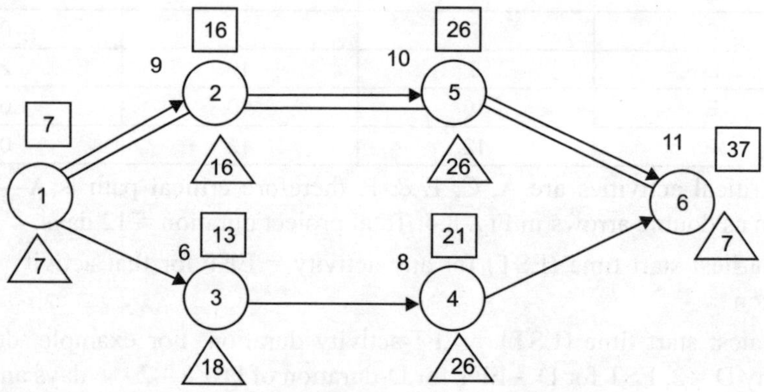

Critical path is the longest path in terms of its duration. If activities on the critical path or critical activities are delayed, the entire project gets delayed. All the resources should be utilized efficiently in order to complete the project on time. There is a slack of 5 on non-critical activities as T.F. is 5 corresponding to them. Even if such activities are delayed, project completion time will not suffer. If need arises, resources may also be transferred from non-critical activities to critical activities. Due to some reasons, if project duration is to be shortened, then duration of critical activities need to be shortened with additional direct cost.

8.1.1 Types of Floats

Total float, *i.e.* (L.F.T.–E.F.T.) is already discussed before. In addition to it, independent float and free float are defined as follows:

(*a*) **Independent Float:** It is the duration by which an activity may be delayed with no effect on the floats of the predecessor activities.

I.F. = EST for succeeding activity – LFT for preceding activity duration.

(*b*) **Free Float:** It is the duration by which an activity may be delayed with no effect on the floats of successor activities.

F.F. for any activity = EST for successor activity – EFT for that activity.

8.2 Programme Evaluation and Review Technique (PERT)

PERT is an event–oriented technique. It is suitable for projects where duration of activities cannot be estimated precisely and there is an uncertainty in activity duration. It is appropriate for probabilistic situations and there are three time estimates for each activity:

(*i*) **Pessimistic Time:** This is based on the pessimistic approach, such as resources are not available as and when required. Obviously this is the maximum time estimate.

(*ii*) **Optimistic Time:** This is based on the optimistic approach, such as all the resources are available as and when required, resources are utilized in the most efficient manner, etc. Obviously this is the least time estimate for completion of an activity.

(*iii*) **The Most Likely Time Estimate:** This is at the intermediate level between pessimistic and optimistic time approach. Its frequency of occurrence is the highest.

Fig. 8.5 shows the Beta distribution which represents the frequency diagram for these time estimates.

T_o = Optimistic time estimate.

T_m = Most likely time estimate.

T_p = Pessimistic time estimate.

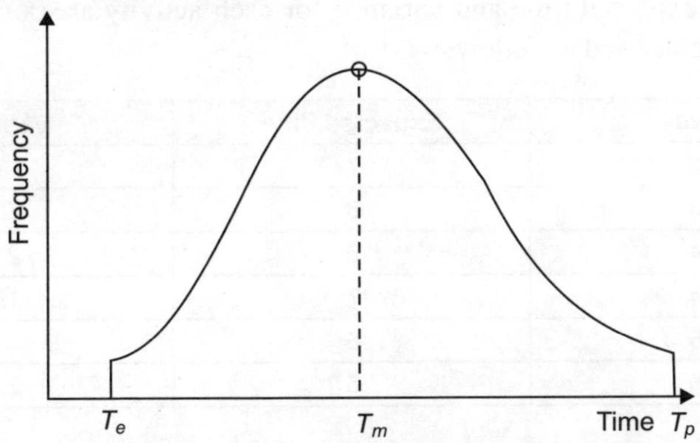

Fig. 8.5 Beta distribution

Variance = $((T_p - To)/6)^2$

Expected Time, $T_e = (T_o + 4T_m + T_p)/6$

From the three time estimates, expected time is calculated as above. In order to obtain the critical path, the rest of the procedure is similar to CPM. Earliest expected time and latest allowable time are obtained for each event as EFT and LFT were obtained in CPM.

Slack = Latest allowable time – Earliest expected time. Slack is zero corresponding to events on critical path.

Example 8.3

Three time estimates are provided for each activity on the following network diagram:

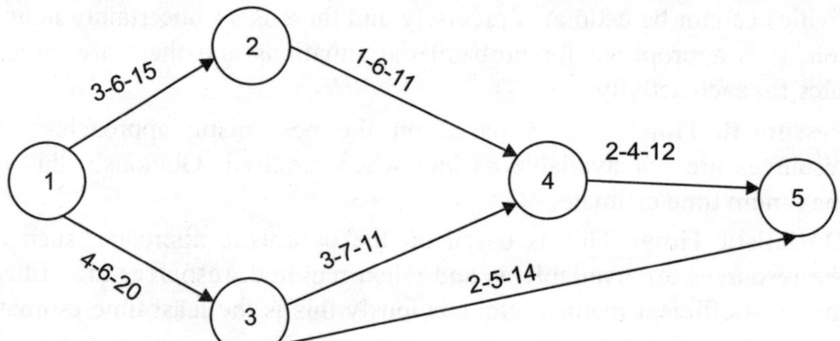

Find out:

(*a*) Expected time for each activity,

(*b*) Variance for each activity,

(*c*) Critical path and total project duration.

As the expected time, $T_e = (T_o + 4T_m + T_p)/6$

For activity 1-2, $T_e = (3 + (4*6) + 15)/6 = 7$

Variance for activity 1-2 = $((15 - 3)/6)^2 = 4$

Similarly expected time and variance for each activity are computed. These results are summarized as follows:

Activity	Expected time	Variance
1-2	7	4
1-3	8	7.11
2-4	6	2.78
3-4	7	1.78
3-5	6	4
4-5	5	2.78

Expected times are given on the network diagram as shown below and critical path is marked:

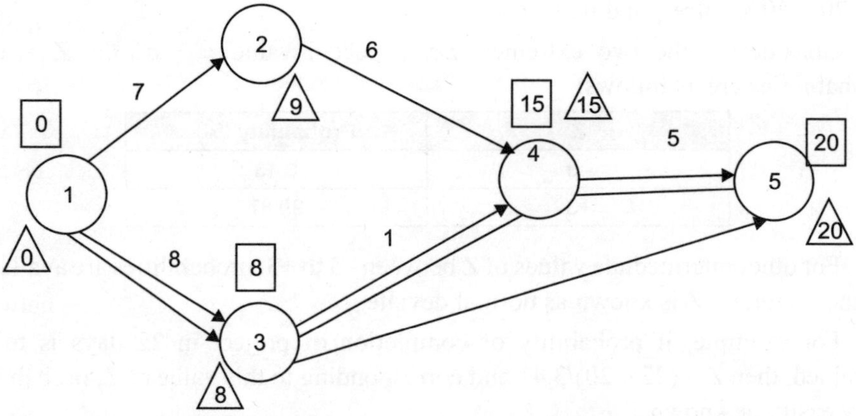

Critical path is 1-3-4-5 with expected total project duration as 20.

Example 8.4

Consider the problem of example 8.3. Calculate the variance of the critical path. What is the probability of completion of project in expected critical path duration as calculated in example 8.3?

Critical activities and their variance are as follows:

Critical activities	Variance
1-3	7.11
3-4	1.78
4-5	2.78

Variance of the critical path = 11.67,.

And standard deviation, $\sigma = \sqrt{(11.67)} = 3.41$.

In order to understand the concept of probability for completion of project, consider the normal distribution as shown below:

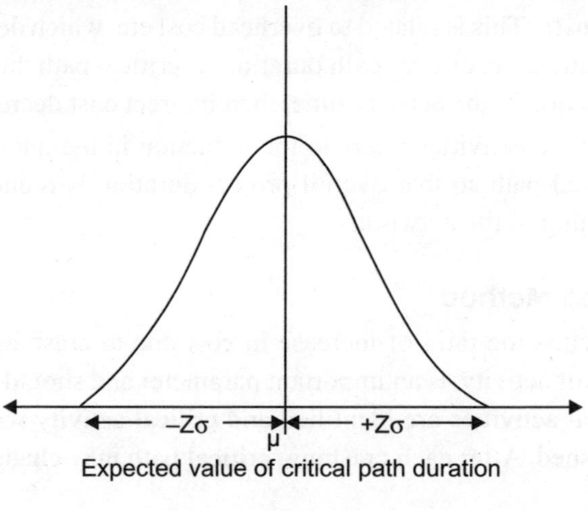

Expected critical path duration or total project duration = 20 which is the central value shown above. Probability of completion of project in this duration, *i.e.* 20 = 50% corresponding to Z = 0.

Considering the two extremes, *i.e.* expected value $\pm 3\ \sigma$, *i.e.* $Z = \pm 3$ probabilities are as follows:

Z	Probability (%)
–3	0.13
+3	99.87

For other intermediate values of Z between –3 to +3, probabilities are available in the literature. Z is known as normal deviate.

For example, if probability of completion of project in 22 days is to be obtained, then Z = (22 – 20)/3.41 and corresponding to this value of Z, probability may easily be known.

$$Z = \frac{\text{Scheduled time against which probability is to be known} - \text{Expected time of completion of project corresponding to critical path}}{\sigma}$$

Probability of completion of project in expected critical path duration is 50% because Z = (20 – 20)/3.41 = 0 and for Z = 0, probability is 50%. With respect to normal distribution, as we move on the positive side of Z, probability is more than 50%. On the negative side of Z, probability is less than 50%.

8.3 Crashing of Activities

Sometimes project needs to be completed earlier than the normal time. This affects project costs.

Total cost of the project consists of two components:

(*i*) **Direct Cost:** These are the costs which can be directly associated with activities. If activity duration is reduced, then resources are required at a higher cost and therefore direct cost increases.

(*ii*) **Indirect Cost:** This is related to overhead cost etc. which depends on overall project duration, *i.e.* critical path duration. If critical path duration is reduced due to reduction in the activity time, then indirect cost decreases.

Crashing of the activities refers to the reduction in the duration of activities on the critical path so that overall project duration is reduced. This is also called crashing of the network.

8.3.1 Crashing Method

Cost slope, which is the ratio of increase in cost due to crashing of activity and decrease in time of activity, is an important parameter and should be computed for activities. Critical activities are identified and critical activity with the minimum cost slope is crashed. After each crashing, critical path may change and should be determined again. The process of crashing the activities is repeated till optimum duration (*i.e.* with minimum total cost) of the project is achieved. If network is to be crashed to a certain limit, then with this method, effect on the cost can also be known.

Example 8.5

Consider the following network diagram:

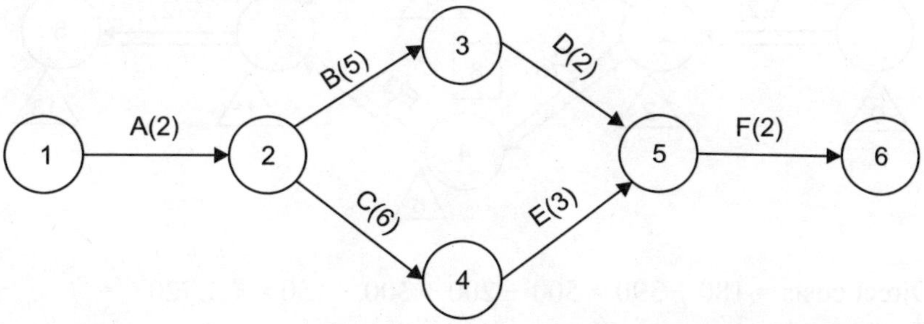

Direct cost along with duration in days for each activity is given as follows:

Activity	Normal		Crash	
	Duration	Cost (₹)	Duration	Cost (₹)
A	2	180	1	280
B	5	390	4	450
C	6	500	3	530
D	2	200	1	400
E	3	·300	2	380
F	2	150	1	300

Assume indirect cost as ₹ 50 per day. By applying crashing method, find out the network with minimum cost.

$$\text{As the cost slope} = \frac{\text{Increase in Cost}}{\text{Decrease in duration}}.$$

$$\text{For C, cost slope} = \frac{530 - 500}{6 - 3} = \frac{30}{3} = ₹\ 10 \text{ per day}$$

Cost slopes of all activities are obtained as follows:

Activity	Cost slope (₹/day)
A	100
B	60
C	10
D	200
E	80
F	150

Critical path is shown on the network:

Total cost without crashing = Direct costs + Indirect cost.

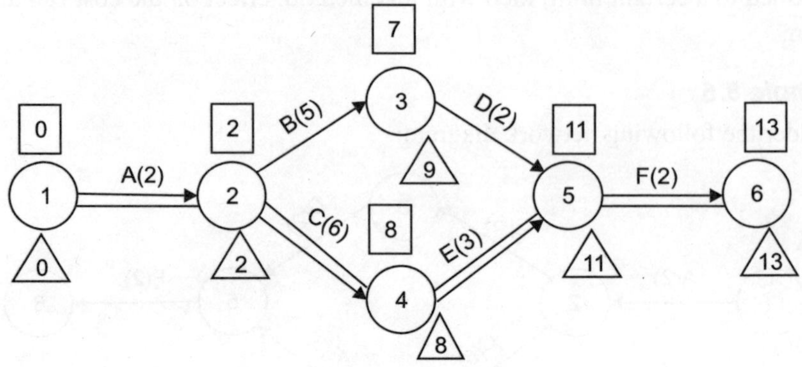

Direct costs = 180 + 390 + 500 + 200 + 300 + 150 = ₹ 1,720.

Indirect cost = 13 × 50 = ₹ 650, as overall project duration is 13 days.

Total Cost = 1,720 + 650 = ₹ 2,370.

Now critical activities are A, C, E and F with cost slope as 100, 180, 150 respectively. Activity C the minimum cost slope, *i.e.* 10; it is to be crashed first.

Crash duration for C is 3 and with the duration, critical path is obtained as follows:

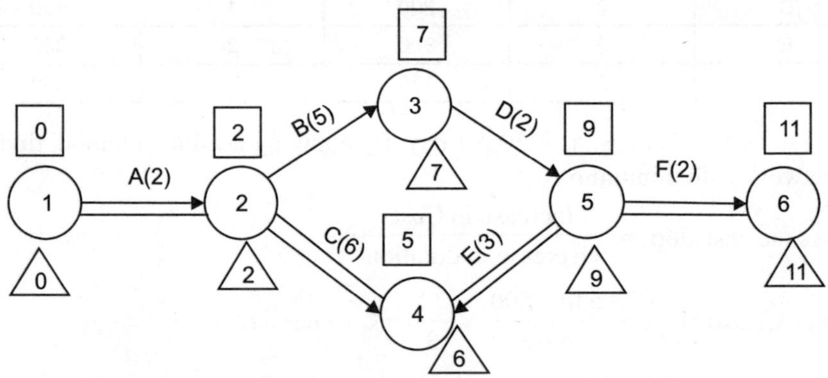

Increase in direct cost due to crashing – 530 – 500 = 30

Now direct cost = 1,720 + 30 = ₹ 1,750.

Indirect cost = 11 × 50 = ₹ 550.

Total Cost = 1,750 + 550 = ₹ 2,300.

Now critical activities are A, B, D and F with cost slopes as 100, 60, 200 and 150 respectively. B is to be crashed now as it is the minimum cost slope:

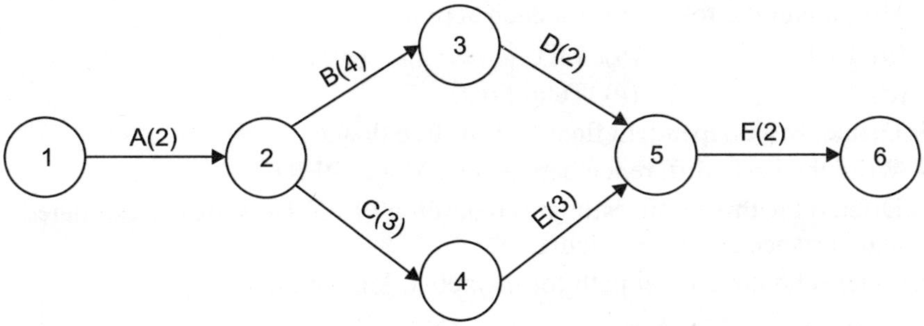

It may be observed that all activities have become critical along with the overall project duration as 10 days. Direct cost will be increased by ₹ 60 and indirect cost will be reduced by ₹ 50.

Direct cost = ₹ 1,810.

Indirect cost = 10 × 50 = ₹ 500.

Total Cost = 1,810 + 500 = ₹ 2,310.

Project duration = 10 days.

Direct cost slope is ₹ 60 per day whereas indirect cost is ₹ 50 per day. As increase in direct cost is more than decrease in indirect cost, total cost has increased from that of the previous stage.

Therefore, minimum cost network is as shown in the previous stage along with the total cost as ₹ 2,300 and project duration is 11 days. However following the same procedure, effect on total cost can be determined by crashing remaining activities along with critical path duration.

EXERCISES

1. What is the limitation of bar chart that is overcome in network methods?
2. Differentiate between activity and event.
3. Explain the dummy activity.
4. Construct the network and determine the critical path considering following details:

Activity	Predecessor	Duration
A	–	3
B	A	1
C	A	3
D	A	2
E	B	6
F	C	1
G	C, D	5
H	E, F, G	2

Also obtain the following for each activity:

(*a*) EFT (*b*) LFT (*c*) EST

(*d*) LST (*e*) Total float.

5. Define: (*a*) independent float and (*b*) free float.

6. Write the basic difference between CPM and PERT.

7. Discuss the three time estimates regarding PERT. How are the estimated time and variance are to be calculated?

8. Determine the critical path for the network diagram:

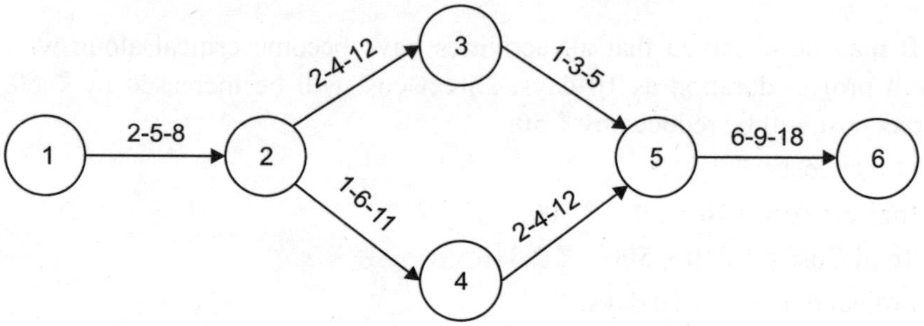

Also obtain the variance for each activity. Calculate the variance of the critical path. Comment on the probability of completing the project in expected critical path duration.

9. What do you understand by crashing of activities? Explain the procedure by taking a suitable example.

❑❑❑

9

Financial Management

In industrial or business organizations where a huge amount of money is involved and stakes are very high, significance of financial management is much more. Financial management deals with proper planning of overall expenditure, how money is to be spent, whether proposal is to be accepted, and priority decision in addition to numerous other functions. Apart from financial management, following are the important functional divisions in an industrial organization:

(*a*) **Manufacturing:** Its aim is to convert raw materials into finished goods of well-defined quality by efficient use of resources. For simplicity, we can include quality control, maintenance, personnel, stores and purchase under the major heading "manufacturing".

(*b*) **Marketing:** Its purpose is sales and distribution of finished goods. In addition to this, marketing division also provides input data for demand forecasting which will be used later to generate the production plan.

(*c*) **Accounts:** In addition to preparation of balance sheets and other routine functions such as record keeping, etc., it is also involved in product cost estimation and process cost accounting.

As we can understand, the integration of various activities always brings fruitful results. Financial management alone cannot function satisfactorily. It depends on other divisions which are 'manufacturing', 'marketing' and 'accounts'.

This interdependence is shown in Fig. 9.1.

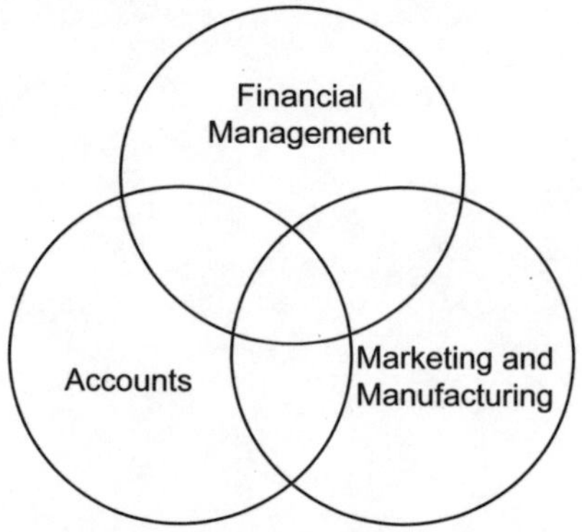

Fig. 9.1 Interdependence of financial management

This will be clearer if we come across following problems which are often encountered:

(*i*) 'Manufacturing' desires that raw material should be in abundance so that production processes never stop for want of material and the resources are not idle. Similarly 'Marketing' requires a huge inventory of finished goods so that these are supplied to the customers as and when desired. At the same time, 'Finance' views inventory as a burden because the capital is blocked for a long period. This is a reason why a trade-off would be required.

(*ii*) Manufacturing is concerned with achievement of production targets satisfactorily. For this purpose as well as for high-quality requirements, newer and sophisticated machines are periodically required. The question is whether it is justified with the present production level.

(*iii*) As time passes, there is wear of the machine parts. Maintenance may not be an easy and cost-effective task. There is a proposal for replacement of the machine. The question is whether it is to be accepted.

(*iv*) An industrial organization is in the process of diversification or is involved in setting up of a new plant. There are alternatives for similar equipment. Which alternative should be selected?

(*v*) As ISO: 9000 organization is dealing in a particular product. Due to stringent quality requirements and total quality management (TQM) approach, emphasis is on production of parts with "zero defects". This firm is not using an old facility for a few months because it generates more number of defects. Suddenly the demand of the parts increases. To take advantage of this situation for profit maximization, should the company not use the old facility since increase in production capacity is the need of the hour?

(*vi*) If a proposal is accepted, then there is a need to invest money. Various sources of finance are as follows:

(*a*) Banks and financial institutions

(*b*) Issuance of bonds

(*c*) Preference/equity shares

Cost of different sources of raising capital is to be estimated and a judicious mix is to be adopted.

(*vii*) By offering the credit to customers for some period, the sales revenue may increase. But a financial analysis needs to be carried out to determine the net effects of credit policy.

(*viii*) Accounting method of the cost has a bearing on financial analysis. Sales price of a product is determined by precise estimation of product cost which in turn depends on process cost accounting among other factors. Further, it is of utmost interest for the management to know the break-even quantity which is the sales volume corresponding to no-profit-no-loss situation.

From the above discussion, it is clear that several proposals/situations queue up before finance professionals for decisions/acceptance. This process is summarized in Fig. 9.2.

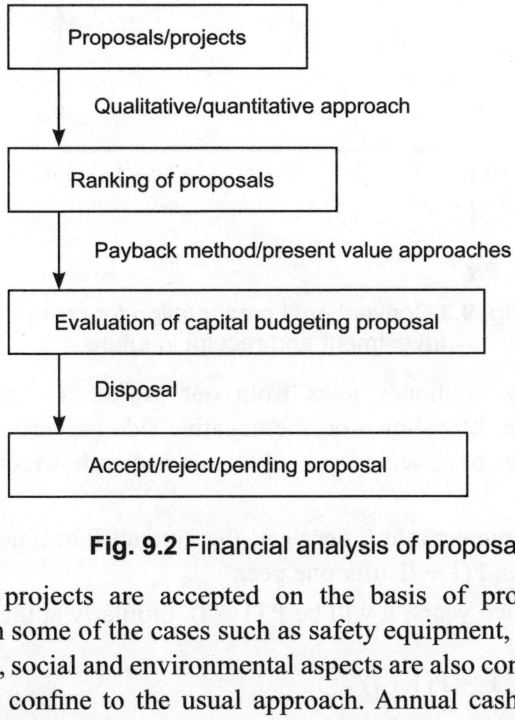

Fig. 9.2 Financial analysis of proposals

Usually the projects are accepted on the basis of profitability by their evaluation. But in some of the cases such as safety equipment, means of arresting the pollution, etc., social and environmental aspects are also considered. However, at present, let us confine to the usual approach. Annual cash flows need to be projected for evaluation of capital budgeting proposals. Determination of these cash flows largely depends on demand forecasting. Forecasting of demands of the products or services is an essential step in financial analysis. A thorough study of economic situations also helps in demand analysis. Macro¬economics and microeconomics have an impact on financial analysis because the environment in which we work affects the decision making. Certain amount of risk is always involved in investment decisions because forecasts involve error. Statistical measures are useful in evaluating the risk such as standard deviation.

Yet another important aspect of financial management is the time value of money. For instance, if a person comes across two proposals: (*i*) Receipt of ₹ 1,000 today, and (*ii*) Receipt of ₹ 1,000 after one year. Obviously, even a layman can tell that the first proposal is preferable because the cash obtained today is more valuable than the similar amount to be received later. Therefore the amount of money alone cannot be analyzed unless time factor is associated with that. Most of the financial analytical studies are primarily centred at present value of future expenditures, receipts, etc.

9.1 Time Value of Money

As discussed before, money in absolute terms cannot be compared. Rather it is to be analyzed in relation to time factor. From our day-to-day experience also, if we deposit any amount P in the bank, then an amount F which is more than P is expected after n number of years. This is shown in Fig. 9.3.

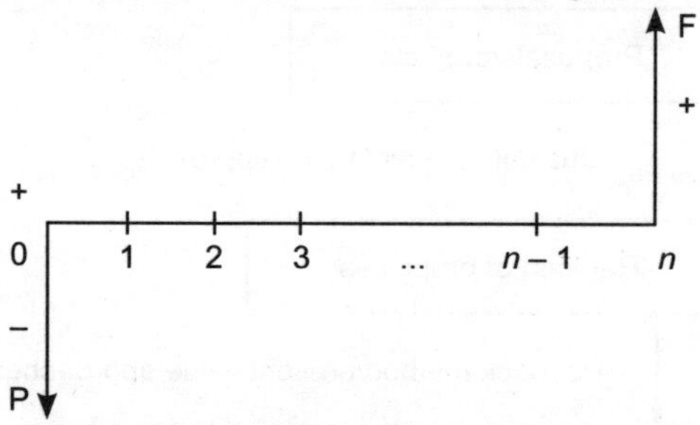

Fig. 9.3 Conventional presentation for present investment and receipt in future

Conventionally, if money goes from our pocket, *i.e.* an expenditure or investment is mode, it is shown on the negative side on vertical axis. Similarly any receipt of money or revenue is a kind of gain and is shown on the positive side in Fig. 9.3.

Assuming an interest rate i, per year, the amount P at time 0, *i.e.* at present becomes $P + Pi$, *i.e.* $P(1 + i)$ after one year.

At the end of two years, it will be $P(1 + i)$. Similarly at the end of n number of years:

$$F = P(1 + i)^n,$$

and
$$P = \frac{f}{(1 + i)^n}, \qquad \text{...(9.1)}$$

where $\frac{1}{(1+i)^n}$ is known as the present value factor. Reciprocal of this, *i.e.* $(1 + i)^n$ is the compound amount factor.

Example 9.1

Work out the present value of future receipt of ₹ 10,000 after 10 years, considering interest rate as 10%. With reference to these data find out (*a*) present value factor (*b*) compound amount factor:

(*a*) Present value factor

From equation (9.1), $P = \dfrac{F}{(1+i)^n}$

Here F= ₹ 10,000, n = 10 years and i = 10% *i.e.* 0.1,

Therefore $P = \dfrac{10,000}{(1.1)^{10}}$ = ₹ 3855.43

(*b*) Compound amount factor = $(1 + i)^n = (1.1)^{10} = 2.5937$

Annuity

There are so many practical situations in which equal amount of money is deposited periodically or received periodically. For example, life insurance policy, in which we deposit equal amount of money yearly/half-yearly/quarterly for a definite period. Similarly bank gives loan to a customer and the customer has to repay the loan yearly in equal amounts for a certain number of years. This kind of situation is shown in Fig 9.4. In this cash flow diagram, *A* is called an annuity which is a series of periodic receipts or payments of equal amount.

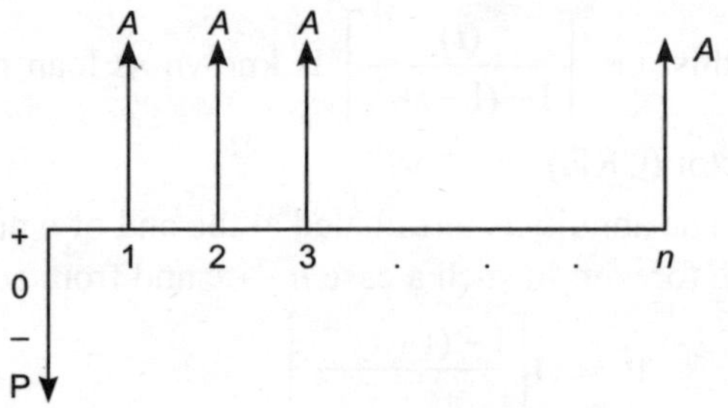

Fig. 9.4 Cash flow diagram

Now from equation (9.1),

Present value corresponding to *A* at the end of year 1

$\dfrac{A}{(1+i)}$ = where I is as usual, *i.e.* interest rate in %.

Similarly P.V. corresponding to *A* at 2 = $\dfrac{A}{(1+i)^2}$

And P.V. for *A* at the end of year $n = \dfrac{A}{(1+i)^n}$.

The present value, P of the annuity is the sum of above values,

$$P = \frac{A}{(1+i)} + \frac{A}{(1+i)^2} + \frac{A}{(1+i)^3} + ... + \frac{A}{(1+i)^n}$$

$$= \frac{A}{(1+i)} \left[1 + \frac{1}{(1+I)} + \frac{1}{(1+i)^2} + ... + \frac{1}{(1+I)^{n-1}}\right].$$

This value in bracket corresponds to well-known geometric progression, $a, ar, \ldots\ldots\ldots, ar^{n-1}$ and sum of this series is:

$$a\frac{(r^n-1)}{(r-1)}.$$

By application of this,

$$P = \frac{A}{(1+i)}\left[\frac{(1+i)^{n-1}}{i\,(1+i)^{n-1}}\right],$$

Or $$P = A\left[\frac{(1+i)^{n-1}}{i(1+i)^n}\right],$$

Or $$P = A\left[\frac{1-(1+i)^{-n}}{(i)}\right]. \qquad \ldots (9.2)$$

Value in bracket, *i.e.* $\left[\frac{1-(1+i)^{-n}}{i}\right]$ is called annuity present value factor.

Reciprocal of this, *i.e.* $\left[\frac{(i)}{1-(1+i)^{-n}}\right]$ is known as loan repayment factor or capital recovery factor (CRF.)

Refer Fig. 9.4. The annuity is terminated at the end of n number of years. But it may also continue forever. In such a case $n \to \infty$ and from equation (9.2),

$$P = A\left[\frac{1-(1+i)^{-00}}{i}\right],$$

Or $$P = A\left[\frac{1-0}{i}\right],$$

Or $$P = \frac{A}{i}.$$

Such type of annuity is known as perpetual annuity. … (9.3)

Example 9.2

An institution is in the process of developing a fund from which gold medals may be given to the toppers every year. It is also estimated that an average expenditure of ₹ 12,000 will be incurred annually. Incidentally an old student of the institution who is a successful profossisal now, wishes to deposit the amount at once for this purpose. How much should he donate to the institution considering an interest rate of 12%, so that medals may be given for a very long time.

Equation (9.3) will be applicable as this concerns perpetual annuity.

Therefore $P = \frac{12000}{0.12}$ = ₹ 1 lakh.

Therefore the old student of the institution should be advised to donate ₹ 1 lakh for this purpose.

Example 9.3

A person decides to deposit ₹ 10,000 at the end of every year in his bank account for 15 years. How much should he receive at the end of 15 years considering annual interest rate as 10%?

This situation is depicted below:

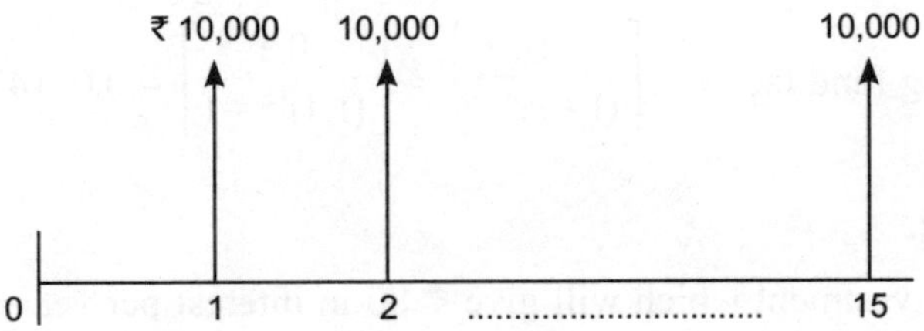

Value at the end of year 1 = 10,000.

Value at the end of year 2 = 10,000 + 10,000 (1+0.1).

Value at the end of year 3 = 10,000 + 10,000 (1.1) + 10,000 $(1.1)^2$.

Value at the end of year 15 = 10,000 + 10,000 (1.1) + 10,000 $(1.1)^2$ +......+ 10,000 $(1.1)^{14}$.

Therefore value at the end of 15 years = 10,000 (1+ (1.1) + $(1.1)^2$ +...+ (($(1.1)^{14}$

$$= (10{,}000 \left[\frac{(1.1)^{15}-1}{(1.1)-1}\right] = ₹\ 3{,}17{,}724.82.$$

The above example can be depicted more generally as shown below:

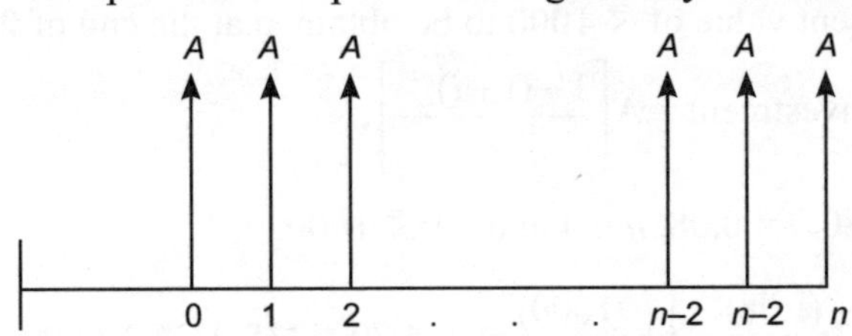

Here Future Value, F = $a + a(1 + i) + a(1 + i)^2 + \ldots.. + a(1 + i)^{n-1}$

$$= (a)\left[\frac{(1+i)^{n-1}}{i}\right],$$

Where $\left[\frac{(1+i)^{n-1}}{i}\right]$ is called as annuity compound amount.

Reciprocal of this, *i.e.* $\left[\frac{i}{(1+i)^{n-1}}\right]$ is know as sinking fund factor.

Familiarity with various factors is useful in financial analysis and planning. Potential applications include valuation of securities, comparison of proposals, and replacement analysis. These are discussed in the subsequent sections. However following problems will help in understanding their use.

Example 9.4

Find out (a) annuity compound amount factor and, (b) sinking fund factor for the data of example 9.3.

$$(a)\text{ Annuity compound amount factor} = \left[\frac{(1+i)^{n-1}}{i}\right] = \left[\frac{(1.1)^{15-1}}{0.1}\right] = 31.7725.$$

$$(b)\text{ Sinking fund factor} = \left[\frac{i}{(1+i)^{n-1}}\right] = \left[\frac{0.1}{(1.1)^{15}-1}\right] = 0.03147.$$

Example 9.5

Consider an investment which will give ₹ 80 in interest per year and ₹ 1,000 at maturity and that is the end of four years from now. What should be the present value of this investment?

Assume interest rate as 8%.

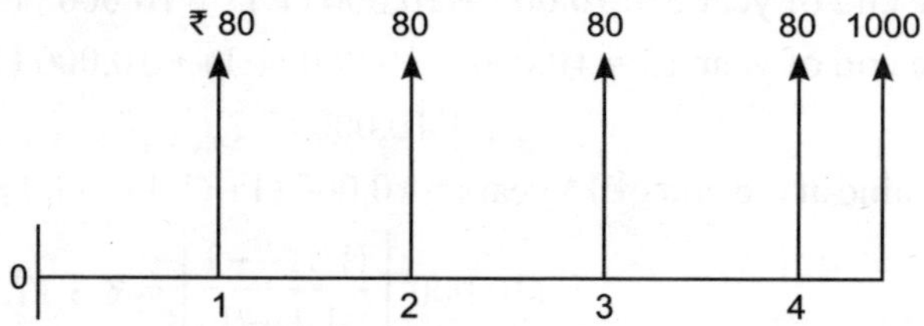

Present value of the investment = present value of the stream of ₹ 80 from 1 to 4 years + present value of ₹ 1000 to be obtained at the end of four years.

$$\text{P.V. of the investment} = A\left[\frac{1-(1+i)^{-n}}{i}\right],$$

Where $A = 80$, $i = 0.08$, $n = 4$ and $F =$ ₹ 1000.

$$\text{P.V.} = 80\left[\frac{1-(1.08)^{-4}}{(0.08)}\right] + \frac{(1000)}{1.08^4} = 264.79 + 735.03 \text{ ₹ } 1{,}000.$$

[**Note:** In this problem, the present value of the investment works out to be the same as maturity value because the coupon interest rate is equal to yield to maturity. At this stage, the readers may note this point. This will be clear in the next section when we discuss valuation of securities.]

9.2 Valuation of Securities

Securities include shares or stocks (equity as well as preference) and bonds among variety of types which are long term as well as short-term. These are issued for raising the capital and are used as sources of finance.

The types of shares and bonds are shown in Fig. 9.5.

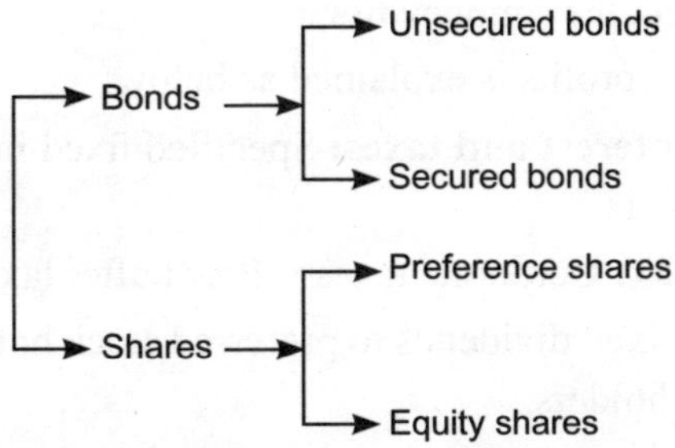

Fig. 9.5 Types of bonds/shares

9.2.1 Bonds

We are familiar with the procedure of getting loan from banks. We pay the interest to the bank and also return the amount of loan at the end of period mutually agreed upon. In a similar manner, industrial organization etc. issue bonds for general public to raise the capital for their projects. The company pays the interest to the bond holder and at the end of the stated number of years, returns the original amount of money also. The bonds are of two types:

(*a*) **Secured bonds:** These are backed by a pledge of specific assets.

(*b*) **Unsecured bonds:** These are also called debentures and this term will be used frequently.

As these are a kind of debt taken by the company, the company is liable to pay interest to the debenture holders whether there is profit or not. Debt/debenture/bond is often used synonymously. Out of the total earnings, first of all debenture holders will get the fixed interest and then taxes are applicable on the remaining amount. In this way, the interest payments are tax deductible. This will be illustrated by a numerical example, which is provided in the next section. After paying taxes, the remaining profit is available to share holders.

9.2.2 Shares

Since shareholders are considered to be owners of the company, their original money is not refunded by the company particularly for equity shares.

(*a*) **Preference shares:** They are also called preferred stocks. As the name suggests, they will get preference over common stocks, *i.e.* dividends will be paid first to preference share holders and then to equity shareholders.

(*b*) **Equity shares:** They are also called common stocks. Equity share holders have voting right. Their risks are high and they have a major say in the company's affairs. They are the last to have a claim over profits.

From the above discussion, it must be understood that risk corresponds to:

Equity shares > Preference shares > Debentures or bonds.

Bond holders are at the least risk because in case of liquidation or winding up of business, they will get their money first.

Distribution of earnings/profits is explained as below:

(*a*) Earning before interest and taxes: Specified fixed interest to bond holders = Earnings befors tax.

(*b*) Earning before tax: Corporate taxes = Profit after tax.

(*c*) Profit after tax: Fixed dividends to preferred stock holders = Profit available to common stock holders.

9.2.3 Security Prices

We often hear news from stock market. There are ups and downs in share prices. Security prices reflect the followings:

(*a*) Confidence of the investor: Investor has made forecasts for the benefit of owning the security and confident of achieving that.

(*b*) Performance of the issuing firm: Past performance of the firm issuing the security and present trend.

(*c*) Overall economic scenario.

(*d*) Risks involved: Normally interest paid by government establishments is considered to be the risk-free return. Then expectations increase as the level of risk increases. This linear relationship is shown in Fig. 9.6.

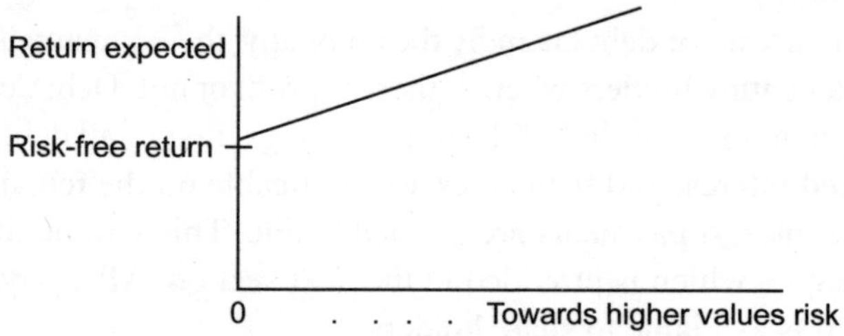

Fig 9.6 Return–risk relationship

(*e*) Present value of the future cash flow: Future cash flow is predicted by the investor and depending on the risk involved, present value is evaluated by discounting at a corresponding rate of return.

9.2.4 Risk

As discussed before the risk are always associated with shares. For example, if the present price of a share is ₹ 100, then after some time, say one year, its price may be higher or lower than ₹ 100 or similar, *i.e.* ₹ 100. If higher, then owner of the security may consider himself to be a gainer. Similarly if the price is lower, then he may think to be at a loss. In such an uncertain environment, one has to estimate the probability of occurrence.

Let us consider the probabilities to be as follows:

S. No.	Price of the share which may be after one year (₹)	Probability
1	70	0.05
2	80	0.1
3	90	0.1
4	100	0.05
5	110	0.2
6	120	0.4
7	130	0.1

It is needless to say that the sum of probabilities of all possible happenings is 1.

To find out the expected value, we have to multiply possible prices with corresponding probabilities and add them.

Expected value for price of share after one year, E(x), =

(70 × 0.5) + (80 × 0.1) + (90 × 0.1) + (100 × 0.05) + (110 × 0.2) + (120 × 0.4) + (130 × 0.1) = ₹ 108.50.

$$\text{Expected rate of return (ROR) in \%} = \frac{(108.50-100)}{(100)} \times 100 = 8.5\%$$

To generalize, if possible values are $X_1, X_2, X_3 \ldots\ldots X_n$ and corresponding probabilities are P_1, P_2, P_3, P_n, then

$$E(x) = P_1X_1 + P_2X_2 + P_3X_3 + \ldots\ldots + P_nX_n \qquad \text{...(9.4)}$$

It may be noted that if probabilities of all possible values are equal, *i.e.*

$$P_1 = P_2 = P_3 = \ldots = P_n = \frac{1}{n}\text{, then}$$

$$E(x) = \frac{X_1 + X_2 + X_3 + \ldots\ldots + X_n}{n} = \frac{\sum X}{n},$$

Which is nothing, but the average value.

In case of an uncertain environment, standard deviation is a measure of risk. It is a measure of the scatter of values.

$$\text{S.D. } (\square) = \sqrt{\sum_{i=1}^{n} \mathrm{P}_i\,[X_i - \mathrm{E}(x)]^2}$$

Considering the present example,

$$\square = \sqrt{\begin{array}{l} 0.05(70-108.50)^2 + 0.1(80-108.50)^2 \\ +0.1(90-108.50)^2 + 0.05(100-108.50)^2 \\ +0.2(110-108.50)^2 + 0.4(120-108.50)^2 \\ +\,0.1\,(130\;-\;108.50)^2 \end{array}}$$

$$= \sqrt{74.1125 + 81.225 + 3225 + 3.6125 + 0.45 + 52.90 + 46.225}\,\cdot$$

If standard deviation is more, risk in investment is more. But this is a suitable measure for evaluating risk associated with individual securities. For comparison of different securities, coefficient of variation is a better measure of risk because it considers standard deviation per unit of expected value.

$$\text{Coefficient of variation} = \frac{\text{Standard deviation}}{\text{Expected value}}$$

$$\textit{i.e.}\ \text{Coefficient of variation} = \frac{\sigma}{E(x)} \quad ...(9.5)$$

In the present example, □ = 17.11 and E(x) = 108.50

$$\text{And coefficient of variation} = \frac{17.11}{108.50} = 0.158$$

To avoid or minimize risk, any investor, would not like to purchase all the shares of a single company or to invest in a single sector. It would be preferable to hold a portfolio of various kinds of securities. This is because if one sector is not performing well and rate of return is negative, another may compensate for it and risk can be minimized. In such case, expected overall rate of return will be sum of multiplications of expected ROR on individual security and proportion of total investment in that security. Obviously the sum of these individual proportions of investment will be unity.

Example 9.6

Expected rate of return for securities 1 to 7 are as follows:

Securities:	1	2	3	4	5	6	7
Expected (ROR)%:	20	8	7	10	15	25	30

Proportion of the investment: 0.1 0.08 0.1 0.07 0.1 0.2 0.35.

Find out the expected rate of return on a portfolio of 7 securities.

In this problem, it may be observed that sum of the proportions of investment is 1. This means that the investor has put 10% of total planned investment in security 1, 8% of that in security 2, 10% in security 3 and so on.

The expected ROR for the portfolio

$$= (20 \times 0.1) + (8 \times 0.08) + (7 \times 0.1) + (10 \times 0.07) + (15 \times 0.1) + (25 \times 0.2) + (30 \times 0.35)$$

$$= 2 + 0.64 + 0.7 + 0.7 + 1.5 + 5 + 10.5 = 21.04\%.$$

The method for calculating the overall expected rate of return is explained in example 9.6. In real practice, some other types of problems may also arise. For instance, an investor desires some minimum expected ROR for the portfolio. At present, he owns some of the securities and wishes to include one or more security so that minimum level of ROR can be achieved. Out of several options, a decision can easily be taken in favour of one.

9.2.5 Intrinsic Value Analysis

This is also known as real worth analysis. If we go to market to purchase any product, we try to assess whether it is worth the price asked. In a similar way, real worth of the security is to be analyzed irrespective of the market price of that at present. While estimating the real worth, one should consider future plans of the company, its growth, its assets and declaration of the firm from time to time. This real worth is to be compared with the proposed price. If real worth is more, then we should procure the stock, otherwise ignore it. This process is also shown in Fig. 9.7.

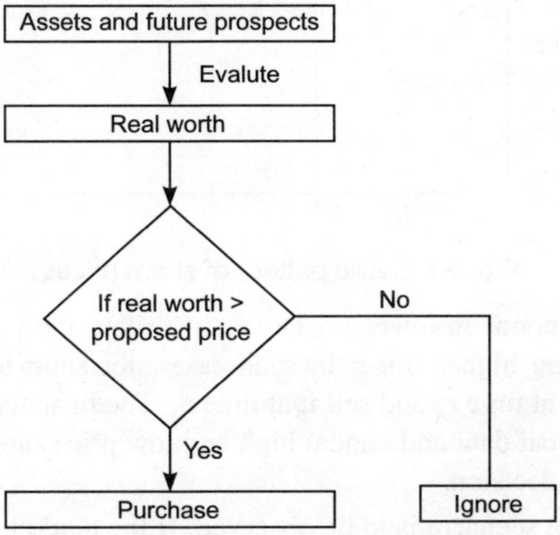

Fig. 9.7 Analysis for purchase! of stocks

As shown in Fig. 9.7, if price of the share is less than its real worth or intrinsic value, we purchase the share in the hope that business world will recognize its real worth after some time and prices will increase and we will gain. But what will happen if share price remains as it is for a very long time, *i.e.* people are unable to recognize its real worth. In that case, our expectations will not come true and own decision may not be fruitful. This is one of the limitations of intrinsic value analysis. Similarly, if the present market price is more than the real worth and we have avoided to purchase it hoping that in near future, prices will come down. This is based on the assumption that market is able to judge the real value soon and it will react accordingly. But market may not act fast in recognizing the real worth of the stock and in such a case, this analysis may not be useful. Future of some of the industries is based on a single event. For example, in case of Oil and Natural Gas Commission (ONGC), the investor may foresee quick growth if oil is found in abundance in some part of the country all of a sudden. In case of Information Technology (IT) firms, physical assets may not be evaluated. Rather its high growth is an issue for valuation.

9.2.6 Current Market Price Approach of Security Valuation

Market prices of securities are observed in this approach and short-term factors are of prime consideration. If we analyze the data for various securities, then we can find out some relationship of stock prices with time. For instance, there are frequent ups & downs in some of the stocks, *i.e.* peak and valley pattern or cyclical pattern. This is shown in Fig. 9.8.

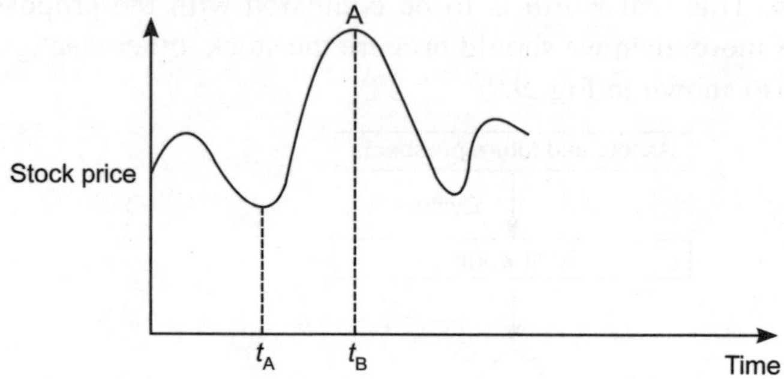

Fig. 9.8 Cyclic pattern of share prices

Point A corresponds to lower prices *i.e.*, at valley, and point B corresponds to peak representing higher price. In such cases, for short-term gain, one can purchase the share at time t_A and sell it at time t_B. The financial analyst will have to study the historical data and annual high and low prices are to be recorded in order to arrive at a decision.

Overall market scenario is to be observed. If the market is not enthusiastic and prices are below the normal expectations for almost all kinds of stocks, then this may be considered as short-lived situation. It may be anticipated that value of the shares may rise again after a short time and stock market will behave in normal way.

The observation reveals that the whole sector is performing well except one particular industry operating in the same business area without any visible reason. Then one can hope that it will also start performing on its own or someone will take over its operations. For example, in case of the software industry, suppose that all the industries' stock values are high except one particular firm XYZ Ltd. Since demands are high, to take advantage of the situation, acquisition of the lagging company is possible and its shares prices may rise. Although stock market is highly speculative in nature, yet this kind of analysis helps in decision making.

As discussed in section 9.1, the present value of future dividends is to be evaluated at a suitable discounting rate. This rate may be required ROR or long-term interest rate because common stocks never mature unlike bonds. One has to predict the future dividends and planning period which may be very large. If A_1, A_2, A_3 .... are future dividends expected at the end of 1, 2, 3......... years and if planning periods are n number of years, then present value:

$$P = \frac{A_1}{(1+i)} + \frac{A_2}{(1+i)^2} + ... + \frac{(A_n)}{(1+i)^n}.$$

If a definite growth pattern can be ascertained, then mathematical modelling and its solution becomes easier.

For instance, a constant growth pattern is shown in Fig 9.9.

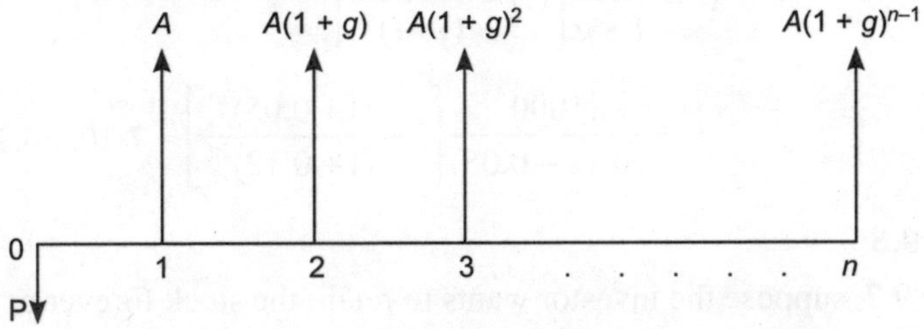

Fig. 9.9 Constant growth pattern of dividends

Dividend is expected to grow at the rate g.

Now $$P = \frac{A}{(1+i)} + \frac{A(1+g)}{(1+i)^2} + \frac{A(1+g)^2}{(1+i)^3} + ... + \frac{A(1+g)^{n-1}}{(1+i)^n}. \qquad ...(9.6)$$

Or $$P = \frac{A}{(1+i)}\left[\frac{1+g)^n}{(1+i)^n}\right]-1.$$

Or $$P = \left[\frac{(1+g)}{(1+i)}\right]-1.$$

Or $$P = \left[\frac{A}{(1+i)^n}\right]\left[\frac{(1+i)^n-(1+g)^n}{(i-g)}\right].$$

Or $$P = \frac{A}{(i-g)}\left[1-\frac{(1+g)^n}{(1+i)^n}\right].$$

If $n \to \infty$ and $i > g$, $\frac{(1+g)^n}{(1+i)^n} \to 0$

And $$P = \frac{A}{(i-g)}. \qquad ...(9.7)$$

Example 9.7

An investor is interested in a common stock which is expected to give a dividend of ₹ 1000 at the end of year 1 and he is planning to retain it for 20 years. The dividend is expected to grow at the rate of 5%. Long-term interest rates are 12% and the investor wants to use it for discounting the future receipts. How should the investor evaluate the present value of this security?

Consider equation (9.6) which is applicable for this case when:

A = ₹ 1000,

n = 20 years,

$$g = 0.05,$$
$$i = 0.12,$$
$$P = \frac{A}{i-g}\left[(1) - \frac{(1+g)^n}{(1+i)^n}\right]$$
$$= \frac{1000}{(0.12-0.05)}\left[1 - \frac{(1+0.05)^{20}}{(1+0.12)^{20}}\right] = ₹\ 10356.30.$$

Example 9.8

In example 9.7, suppose the investor wants to retain the stock forever or for a very long time. Find out the approximate present worth.

Equation (9.7) will be suitable to use as $n \to \infty$ and $i > g$,

$$\text{therefore P} = \frac{A}{(i-g)} = \frac{1000}{(0.12-0.05)} = ₹\ 14285.74,$$

9.2.7 Trading of Bonds

As discussed in the beginning of this section, the company issuing bonds will pay the interest periodically to the bond holder at the end of the stated period. Let us consider a situation which is shown below:

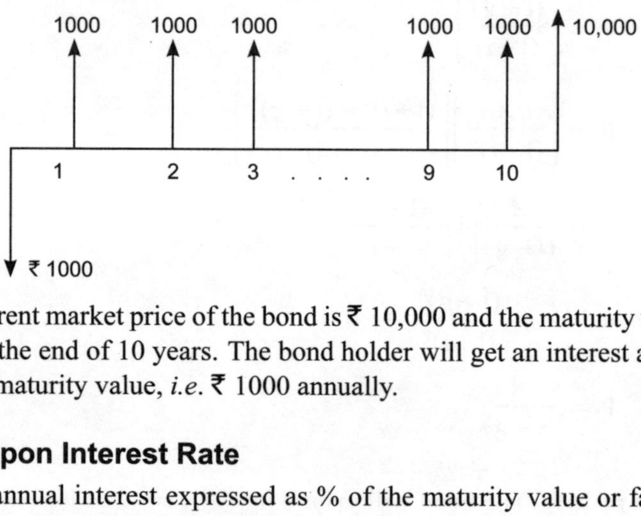

The current market price of the bond is ₹ 10,000 and the maturity value is also ₹ 10,000 at the end of 10 years. The bond holder will get an interest at the rate of 10% of the maturity value, *i.e.* ₹ 1000 annually.

9.2.8 Coupon Interest Rate

This is the annual interest expressed as % of the maturity value or face value of the bond. Now consider equation (9.1) and (9.2), the present value of all future receipts:

$$P = \frac{F}{(1+i)^n} + A\left[\frac{1-(1+i)^{-n}}{i}\right] \quad \text{... (9.8)}$$

In this example, P = ₹ 10,000,

F = ₹ 10,000,

And A = ₹ 1,000,

and
$$10{,}000 = \frac{10{,}000}{(1+i)-10} + 1000\left[\frac{1-(1+i)^{-10}}{i}\right] \quad \text{... (9.9)}$$

9.2.9 Yield to Maturity

This is the rate by which the present investment yields into future benefits. Consider equation (9.9), the value of i is the yield to maturity. If we put $i = 10\% = 0.1$, then the equation is satisfied. In this example, the yield to maturity is equal to coupon interest rate and this bond is considered to be trading at par.

If the present market price is more than the maturity value, *i.e.* $P > F$ or $P >$ ₹ 10,000 and if we formulate the equation (9.9) accordingly, then value of i to satisfy that equation will be less than 10%. In other words, the yield to maturity is less than coupon interest rate. The bond is called to be trading at premium because current price of the bond is more than its maturity value.

On the contrary if the value of P is less than F, then i which will satisfy the equation (9.9) thus formulated is more than 10%, *i.e.* yield to maturity is more than the coupon interest rate. Since the price of the bond paid by customer is less than its maturity value, the bond is said to be trading at discount.

Revisit example 9.5 in which present value of the investment was calculated from fundamentals which worked out to be similar to maturity value. This was because yield to maturity was equal to coupon interest rate and trading was at par.

Example 9.9

Consider example 9.5 in which present value and maturity values are equal, *i.e.* ₹ 1000 and investor will receive ₹ 80 for four years. Now suppose that after the issuance of security, the interest rate dropped to 6%, how will the present value of the investment be estimated?

In this problem, since the interest rate has changed, any investor would like to discount the future payments with the prevailing rate of interest.

$$\text{Therefore, } P = 80\left[\frac{1-(1.06)^4}{0.06}\right] + \frac{1000}{(1.06)^4} = ₹\ 1069.30.$$

In the changed economic condition, the present worth has increased.

Example 9.10

The bonds of a company have been sold in multiples of ₹ 5,000 at a coupon interest rate of 12%. The bond will mature in five years and interest will be paid twice in a year, *i.e.* semiannually. Two years later the annual interest rate becomes 10%. Due to some reasons, the bond holder wants to resell the security. How much worth will one bond be?

This problem corresponds to secondary market, *i.e.* the trading of securities which have been issued earlier. On the other hand, primary market relates to the securities which are being issued for the first time.

Original stream of cash flow for one bond is shown below:

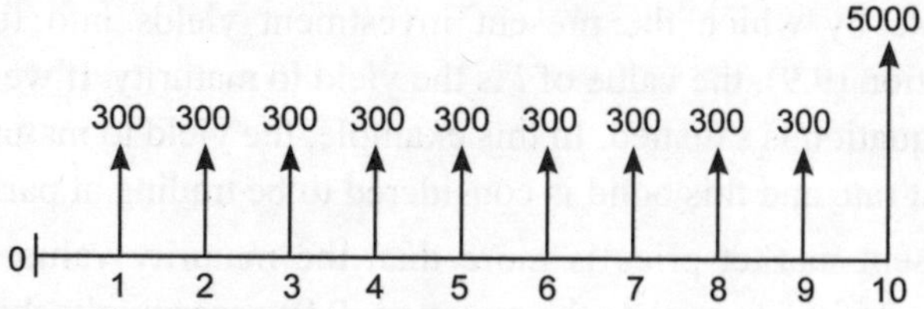

Annual interest = 5000 × 0.12 = ₹ 600.

Since the interest is to be paid half-yearly, the figure above shows 10 half-years, *i.e.* total duration of 5 years which is maturity period.

$$\text{Half-yearly interest} = \frac{600}{2} \text{ ₹ } 300.$$

After two years, *i.e.* when bond holder already received interest for 4 periods, annual interest rates reduced to 10%. Now the present worth of the bond will correspond to cash flow shown below:

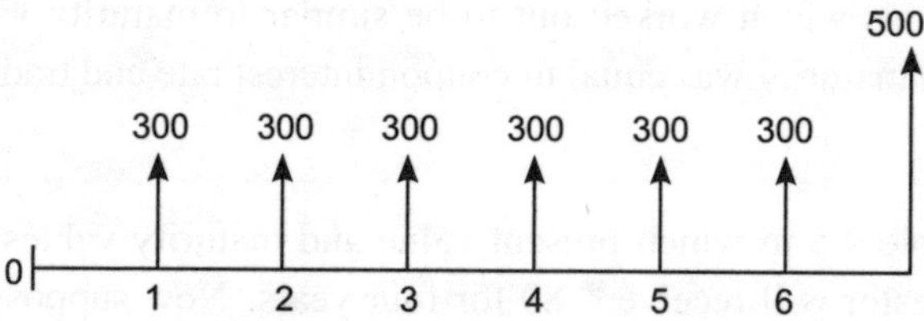

This shows the cash stream for the remaining 3 years, *i.e.* for 6 periods.

$$\text{Now the real worth of one bond} = (300)\left[\frac{1-(1.05)^{-6}}{(0.05)}\right] + \frac{(5000)}{(1.1)^3}$$

$$= 1522.71 + 3756.57$$

$$= \text{₹ } 5279.28$$

9.3 Capital Structure

Business organization are engaged in production, exploration, projects etc. For these activities, they require finance. For raising the finance, the companies issue securities such as shares, bonds etc. All the sources of finance have their advantages and disadvantages. For instance, in case of debt *i.e.*, bond/debenture, the interest payments are tax deductible. This can be understood from the following example:

Suppose that a company needs ₹ 0,000 for its project (Note: to keep the illustration simple, a very small figure is considered. In practice, it may be a very large amount). Assume that the firm raises the entire amount in the form of debt by issuing bonds and promises to pay the bond holders interest at the rate of 10%. Consider the expected annual earning from the project as ₹ 5000 and tax rate as 30%.

The calculations for projected income are shown in Table 9.1.

Table 9.1 Projected income with bonds as source of finance

Annual earning	₹ 5000
Interest to bond holders	₹ 1000
Taxable income	₹ 4000
Tax @ 30%	₹ 1200
Projected income	₹ 2800

Since the company has raised ₹ 10,000 in the form of debt, it is liable to pay interest @10%. This is why interest of ₹ 1000, *i.e.* 10% of ₹ 10,000 is deducted from the earning to get the taxable income. Now we can easily understand that interest payments are tax-deductible. This is the main advantage of debt. On the other hand, the firm has to pay the interest even if no or less earning is there due to some reasons.

Now let us consider the preference shares as the only source of finance in the present example. Fixed dividends are to be paid to share holders. Let us say @15%. The calculations are shown in Table 9.2.

Table 9.2 Projected income with preference shares as source of finance

Earning	₹ 5000
Interest	Nil
Taxable income	₹ 5000
Tax @ 30%	₹ 1500
Earning after tax	₹ 3500
Dividends to preference share holders	₹ 1500
Projected income	₹ 2000

Since only preference shares were issued, no interest payment is there and the company could not take advantage of reduced taxable income. Dividends are 15% of 10,000, *i.e.* ₹ 1500 and projected income is ₹ 2000 which is less than that shown in Table 9.1.

Table 9.3 shows the projected income with equity shares as the only source of finance. Assume that equity share holders desire 17.5% rate of return.

Table 9.3 Projected income with equity shares as source of finance

Earning	₹ 5000
Interest	Nil
Taxable income	₹ 5000
Tax @ 30%	₹ 1500
Earning after tax	₹ 3500
Dividends to preference share holders	Nil
Projected income to common stock holders	₹ 3500

Suppose that 1000 equity shares have been issued out of which 500 were held by original promoters.

Now earning per share (EPS)

$$= \frac{\text{Earning to common stock holders}}{\text{Total number of equity shares}} = \frac{3500}{1000} = 3.5$$

In this example, it was deliberately stated that the required ROR for equity shares is 17.5%. Because after distributing ₹ 1750 (which is also evident from multiplication of EPS, *i.e.* 3.5 and 500 shares purchased by others), remaining ₹ 1750 will be available to the company as income. This was a simple example. In reality, a lot of uncertainty is there in equity shares. If earnings are less, EPS will be less and shareholders may get less than required ROR. On the other hand, in case of high earnings, share holders will get much more than required ROR since EPS is high. However investor expects minimum required ROR and it depends on the risk involved. The company raising the finance may view it as cost of capital, if capital is gained from equity shares.

9.3.1 Cost of Capital

This is the minimum required rate of return. Suppose a person takes loan of ₹ 1 lakh from a bank and he has to pay interest @ 15%, *i.e.* ₹ 15000 per year. Then he should be able to invest the money in such a way that earning should be minimum ₹ 15000 so that liability can be handled. If the earnings are more, then remainder will be profit to him. In other words, 15% is the cost of capital.

Similarly industrial or business organizations raise finance from different sources and costs of different sources of raising capital are as follows:

(*a*) Cost of debt: In the example, it is 10%, *i.e.* the rate of interest paid to bondholders

(*b*) Cost of preference shares: In the example, it is 15%, *i.e.* the rate at which fixed dividend is to be paid to preference share holders

(*c*) Cost of equity shares: This is the reflection of minimum required rate of return by common stock holders or current pattern of ROR on equity shares depending on general economic environment. In the example, it is considered to be 17.5%

In addition to these three sources, sometimes a decision may be taken by the management to retain the earnings for future investment. Suppose that industry requires immediate investment. In stead of issuing new securities which will involve time, effort, as well as cost, it does not distribute the profits to shareholders and retains the profits for further use. Prima facie it may not look good from investors' point of view. But in many cases, they may welcome the move. Because, if investor is not in urgent need of money and he receives the dividends from company, the problem of reinvesting the money will be faced. Further he may have to pay the taxes over the dividends obtained in that financial year. At the same time, the implications for the management of industry lies in better performance because expectations of the investor at a later date will be more. However cost of retained earnings is difficult to estimate.

As discussed in section 9.2, the risk correspond to

Equity shares > Preferred stocks > Debt.

As the risk increases, the required rate of return increases. This is the reason why 17.5%, 15% and 10% cost of capital are considered for these sources of raising capital respectively in the example.

9.3.2 Weighted Average Cost of Capital

Individual sources of raising capital, *i.e.* debt, preference shares and equity shares have been considered in the example before. Each source has advantages and disadvantages. In practice, the capital structure consists of suitable proportions of debt, preference shares and equity shares.

If k_d is before tax cost of debt, then

Effective cost of debt = $k_d(1 - T)$ where T is the tax rate applicable. Weighted average cost of capital,

$$K_{wa} = f_d.k_d(1 - T) + f_p.k_p + f_e.k_e \quad \text{...(9.10)}$$

Where f_d = Proportion of debt in capital structure,

f_p = Proportion of preferred stocks in capital structure,

f_e = Proportion of equity stocks in capital structure,

k_p = Cost of capital concerning preference shares,

k_e = Required return on equity shares.

The weighted average cost of capital, K_{wa}, can also be called alternatively the weighted average required return, because this is the minimum required overall ROR which the business must achieve in order to satisfy the security holders.

Example 9.11

The capital structure of an enterprise consists of:

(*a*) Debt of ₹ 2 lakhs at 10% interest

(*b*) Preference shares of ₹ 1 lakh at 15% dividend

(*c*) 90,000 number of equity shares have been issued @ ₹ 10.

If the current level of ROR on equity shares is 17.5% and tax applicable to the enterprise is @ 30%, then calculate the weighted average required return.

In this problem, first we will have to calculate the total value of the capital, which is equal to

2 lakhs + 1 lakh + (90,000 × 10) = ₹ 12 lakh

Consider equation (9.10),

$$K_{wa} = f_d k_d(1 - T) + f_p.k_p + f_e k_e.$$

Here $f_d = \dfrac{2}{12}$,

$f_p = \dfrac{1}{12}$,

$$f_e = \frac{9}{12},$$
$$k_d = 0.1,$$
$$k_p = 0.15,$$
$$k_e = 0.175,$$
And $$T = 0.3.$$

Substituting these values,

$$K_{wa} = \left[\frac{2}{12}\right] \times 0.1(1-0.3) + \left[\frac{1}{12}\right] \times 0.15 + \left[\frac{9}{12}\right] \times 0.175 = 0.1554.$$

Therefore 15.54% is the weighted average required return.

In Tables 9.1, 9.2 and 9.3, we have referred to earning as ₹ 5000. In financial terms, it is 'earnings before interest and taxes' (EBIT). This is the earning before any tax is deducted and interest is paid to bond holders.

As discussed before, $\text{EPS} = \dfrac{\text{Earning to common shareholders}}{\text{Total number of equity shares}}$

Consider calculation of Table 9.3:

$$\text{EPS} = \frac{\text{Earning after tax} - \text{Dividends to preferred stock holders}}{\text{Total number of equity shares}}$$

$$\text{Or EPS} = \frac{\text{(Taxable income) (Tax rate} \quad \text{Taxable income)} - \text{Dividends to preferred stock holders}}{\text{Total number of equity shares}}$$

$$\text{Or EPS} = \frac{\text{(Taxable income)(1-Tax rate)} - \text{Dividends to preferred stock holders}}{\text{Total number of equity shares}}$$

As taxable income = EBIT, Interest, (EBIT – Interest)(1 – Tax rate)

$$\text{EPS} = \frac{\text{Dividends to preferred stocks holders}}{\text{Total number of equity shares}}$$

If the capital structure, *i.e.* suitable proportion of securities issued by firm is predetermined and tax rate is fixed, then variable terms in the above equation are EPS and EBIT. In such a case, graphic relationship between EPS and EBIT will be a straight line as the above equation resembles the familiar $y = a + bx$. This is shown in Fig. 9.9.

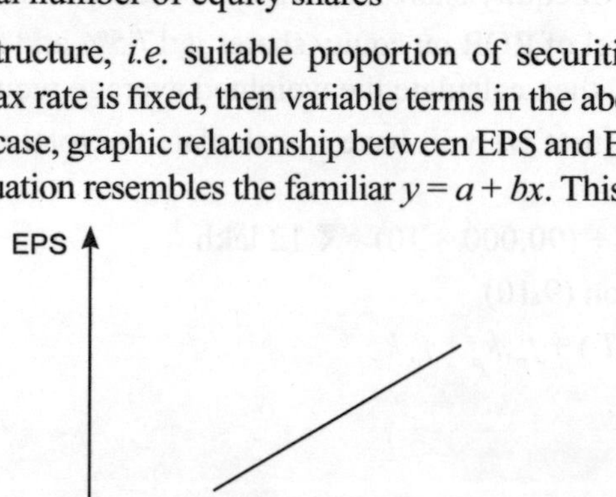

Fig. 9.9 Relationship between EPS and EBIT

For a particular capital structure, to draw this linear relationship, we need two points (EPS 1, EBIT 1) and (EPS 2, EBIT 2). For two different values of EBIT, one can find out the corresponding EPS and draw the relationship.

9.3.3 Optimal Capital Structure

The prime objective of any business organization is profitability. Maximizing the value of the firm in terms of stock value or EPS is the main aim of an efficient capital structure. Optimal capital structure is a set of proportions of various kinds of securities issued by the enterprise to achieve this objective.

For various types of capital structure, the analyst may draw the relationship between EPS & EBIT. From this, he can identify that for which range of expected EBIT, which type of capital structure would be suitable in order to have the maximum possible EPS.

9.3.4 Degree of Financial Leverage

This is in order to determine how sensitive EPS is with respect to change in EBIT. It is in one way an indicator how efficiently the company borrows funds.

$$\text{Degree of financial leverage} = \frac{\%\text{ change in EPS}}{\%\text{ change in EBIT}}.$$

9.3.5 Degree of Operating Leverage

It is an indicator of the sensitivity of EBIT with respect to change in quantity produced and sold. This is also a reflection of efficiencies of production and marketing functions of the organisation.

$$\text{Degree of operating leverage} = \frac{\%\text{ change in EBIT}}{\%\text{ change in quantity sold}}.$$

9.4 Capital Budgeting

Whether it is conversion of raw materials in finished goods as shown in Fig. 9.10 or providing services, plant and equipment are needed in one form or another. These will be utilized throughout their life span which is usually many years. These are known as capital assets.

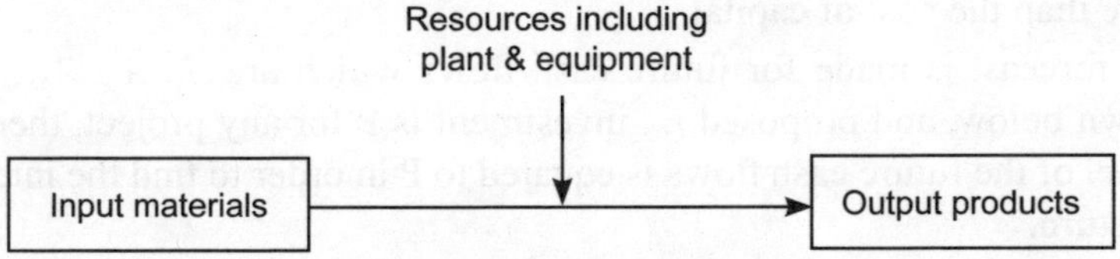

Fig. 9.10 Conversion of input materials into output

We know that budgeting is concerned with knowing the finances available and how to spend it. Similarly capital budgeting is associated with:

(*a*) What kind of capital assets are to be procured?

(*b*) How much expenditure is allowed for procuring these capital assets?

9.4.1 Capital Budgeting Proposals and their Evaluation

Variety of proposals wait for financial approval. These may include:

(*a*) **Procurement and Installation of a New Machine/group of Machines:** Suppose that the existing product or service is not profitable enough and the company wants to diversify into new area. For this purpose it needs some new machines. Also the existing machines are worn out and are not producing quality product whose market is attractive for the company. For this purpose as well as to be productive enough, new equipment are required and they are proposed to replace old equipment.

(*b*) **R & D Projects:** To develop improved design of a product–value added products, some computers along with recent software packages are required. Similarly for the modification of the production processes to become cost-effective, some of the developmental activities require appropriate finances.

(*c*) **Capacity Enhancement:** This may be desirable due to reasons such as (*i*) great demand of the product, (*ii*) customer is willing to pay more price for the product which is also short in supply, etc.

Various proposals are evaluated by the financial analysts. Essential input requirement for this is the forecasts of annual earnings/cash flow/costs. Different methods/approaches used for evaluation are as follows:

1. **Payback Period Method:** Suppose we invest ₹ 10,000 in procuring an equipment and expected annual earning due to that is ₹ 2000. We can easily say that costs involved will be recoverd in 5 years. In this process, we have divided capital investment by expected annual earning. This is called as payback period.

$$\text{Payback period in years} = \frac{\text{Capital investment}}{\text{Expected annual return}}$$

This is the simplest procedure of evaluating various projects/proposals. Everyone wants higher annual return. Higher the annual earnings, lower the payback period which is desirable.

2. **Internal Rate of Return Approach:** The cost of capital which is the minimum required rate of return was discussed in the previous section. In order to earn profit, the project or proposal must earn rate of return which is more than the cost of capital.

If a forecast is made for future cash flows which are A_1, A_2, A_3, ..., An as shown below, and proposed *i*... investment is P for any project, then present value. of the future cash flows is equated to P in order to find the internal rate of return.

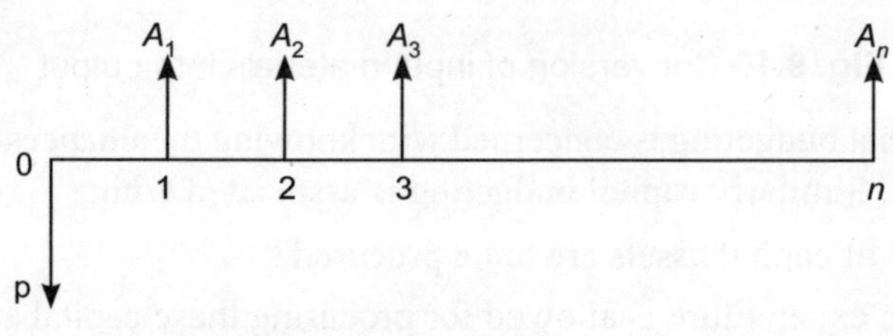

$$P = \frac{A_1}{1+r} + \frac{A_2}{(1+r)^2} + \frac{A_3}{(1+r)^3} + ... \frac{A_n}{(1+r)^n} \quad ...(9.11)$$

The value of r is found from the above equation. With the help of computer, one can start from a random value of r and then search is started for the exact value of r which will satisfy the equation. This way, iteratively the internal rate of return (IRR), *i.e.* r is found.

The analyst can expect the project to be profitable if r is greater than the cost of capital. If r is less than the cost of capital, then the proposal may be rejected.

3. **NPV Method:** The fundamental principle behind the net present value (NPV) method is similar to IRR approach. Since computational difficulties as discussed in IRR approach are avoided in the NPV method, this method is the preferred one.

Net present value

$$P = \frac{A_1}{1+r} + \frac{A_2}{(1+r)^2} + \frac{A_3}{(1+r)^3} + ... \frac{A_n}{(1+r)^n} - P.$$

In case of annuity, NPV

$$= \frac{A_1}{1+r} + \frac{A_2}{(1+r)^2} + ... + \frac{A_n}{(1+r)^n} - P \cdot$$

$$NPV = A\left[\frac{1-(1+r)^{-n}}{r}\right] - P \quad ... (9.12)$$

Substitute the cost of capital or the required return as the value of r in equation (9.12) and find out the NPV. If the NPV is positive, then the benefits are more than the investment and the project is accepted as it yields profits.

The proposal should not be accepted if the NPV is negative as it results in losses. We are at the point of indifference if NPV is zero.

Example 9.12

An entrepreneur is considering the procurement of metallizing equipment at the inital investment of ₹ 1,00,000. Its life is 12 years and projected cash flows per year are approximately ₹ 30,000 for the life of the equipment. If the cost of capital is 10%, state whether the entrepreneur should decide in favour of procurement, on the basis of NPV method.

As we have to find out the NPV for the proposal,

$$NPV = A\left[\frac{1-(1+r)^{-n}}{r}\right] - P$$

Here A = ₹ 30,000

$$R = 12 \text{ years},$$
$$N = 12 \text{ years},$$

And $$P = ₹\ 100{,}000.$$

Therefore $$NPV = 30{,}000\left[\frac{1-(1.1)^{-12}}{0.1}\right] - 100{,}000$$
$$= 204401.75 - 100{,}000 = ₹\ 104410.75.$$

As the NPV is positive, the entrepreneur must decide in favour of procurement of the equipment.

Example 9.13

The purchase of mechanical press is being planned by a firm which costs ₹ 1 lakh. Assume that the scrap value of the press after five years, useful life is zero. The expected cash flows for five years are as follows:

₹ 40000, 40000, 40000, 30000, and 20000.

Find out:

(*a*) Payback period

(*b*) NPV, if cost of capital is 12%

The situation is shown in the figure below:

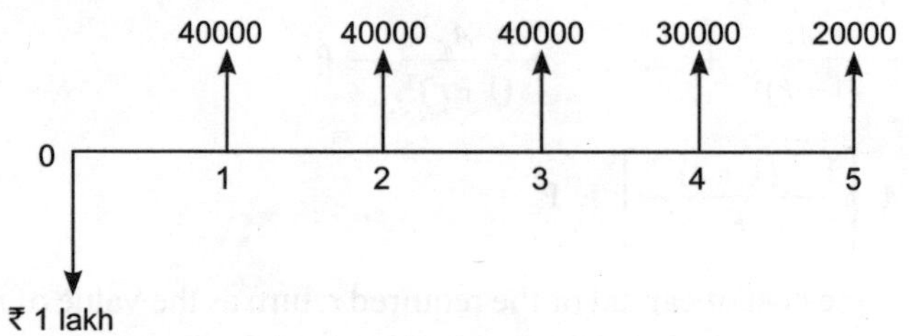

(*a*) The payback period is the ratio of capital investment and expected annual return. But in this fundamental definition, it is assumed that the expected annual return or cash flows will be uniform. In order words, they will form an annuity. This is not the case in present example. Therefore one has to observe the cash flows and conclude that capital investment will be recovered within three years. This is because total of cash flows at the end of three years is 40000 + 40000 + 40000 = 120,000 which exceeds ₹ 1 lakh whereas at the end of two years it is only 80000, *i.e.* less than ₹ 1 lakh. Now consider the average of cash flows for three years which is ₹ 40,000 and payback period:

$$\frac{100{,}000}{40{,}000} = 25 \text{ years.}$$

The drawback of payback period is that it does not consider the time value of money. Further it is difficult to accept or reject the project on the basis of payback period alone. However it gives an idea for the period in which the initial investment will be recovered and whether the management can allow for that period.

(*b*) Refer the figure and as:

$$(\text{NPV}) = \left[\frac{A_1}{(1+r)}\right] + \left[\frac{A_2}{(1+r)^2}\right] + \ldots + \left[\frac{A_n}{(1+r)^n}\right] - \text{P},$$

Where $A_1 = A_2 = A_3 = ₹\ 40{,}000,$

$A_4 = ₹\ 30{,}000,$

$A_5 = ₹\ 20{,}000,$

$n = 5,$

$\text{P} = ₹\ 100{,}000,$

$r = 0.12.$

The NPV is positive, the mechanical press can be purchased.

9.4.2 Net Cash Flow

Broadly speaking the cash flow is annual revenue minus yearly expenditure. But the net cash flow is net of taxes, *i.e.* taxes are deducted from the cash flow. In order to find out the taxable income, depreciation is reduced from revenue – expenditure. In other words, depreciation is tax–deductible.

Depreciation is the reduction in value of the equipment as time passes. For example, consider the present value of an equipment as ₹ 14000 and it is expected that at the end of 12 years, useful life, its salvage value will be ₹ 2000. Then overall depreciation for the life of equipment is (14000 – 2000), *i.e.* ₹ 12000 and yearly depreciation may be considered as 12000/12 = ₹ 1000. There are more than one methods of calculating the depreciation which are beyond the scope. The firm may depreciate the equipment completely over some numbers of years which are less than the life of equipment.

Table 9.4 shows the calculations for net cash flow where depreciation per year for the plant and equipment of the company is ₹ 100,000. Yearly revenue and expenses are ₹ 10 lakh and 4 lakh respectively. Tax rate is assumed to be 35%.

Table 9.4 Calculation of net cash flow

Revenue (A)	₹ 10,000,00
Expenses (B)	₹ 4,000,00
(A–B)	₹ 600,000
Less depreciation	₹ 100,000
Taxable income	₹ 500,000
Tax @ 35%	₹ 175,000
Net cash flow (A–B–Tax)	₹ 425,000 (600,000 – 175,000)

Example 9.14

Consider the data shown in Table 9.4 for a company, a proposal of purchasing new metallizing equipment is pending for the financial analysis whose cost is ₹ 100,000. The proposed equipment will improve the operations in terms of cost reduction, *i.e.* expenses will be reduced by ₹ 40,000 per year. Assume that annual revenue is unaffected. The equipment is to be depreciated over 8 years. Life of the equipment is 12 years. If the cost of capital is 10%, compute the NPV for the proposal and make a decision whether the proposal is to be accepted.

$$\text{Yearly depreciation for the proposed equipment} = \frac{100,000}{8} = ₹\ 12500.$$

It is of relevance to consider the net effect on the cash flow due to purchase of the proposed equipment. The annual depreciation of ₹ 12500 will be applicable for the first 8 years. As the life of the equipment is 12 years, for the remaining 4 years, depreciation will not be considered. Table 9.5 shows the change in cash flow for the first 8 years.

Table 9.5 Change in cash flow for the first 8 years

Change in revenue (A)	0
Decrease in expenses (B)	– 40,000
(A–B)	₹ 40,000
Less depreciation	₹ 12,500
Change in taxable income	₹ 27,500
Tax @ 35%	₹ 9,625
Increased cash flow (A–B–Tax)	₹ 30,375

₹ 30375 is the increased cash flow for the first 8 years due to purchase of the equipment. For the remaining 4 years, similar type of calculations can be made except depreciation will be zero for the proposed addition of equipment. This is shown in Table 9.6.

Table 9.6 Increase in cash flow for 9–12 Years

Change in revenue (A)	0
Decrease in expenses (B)	–40,000
(A–B)	₹ 40,000
Less depreciation	0
Change in taxable income	₹ 40,000
Tax @ 35%	₹ 14,000
Increased cash flow (A–B–Tax)	₹ 26,000

The increased cash flow diagram is shown in Fig. 9.11 where:

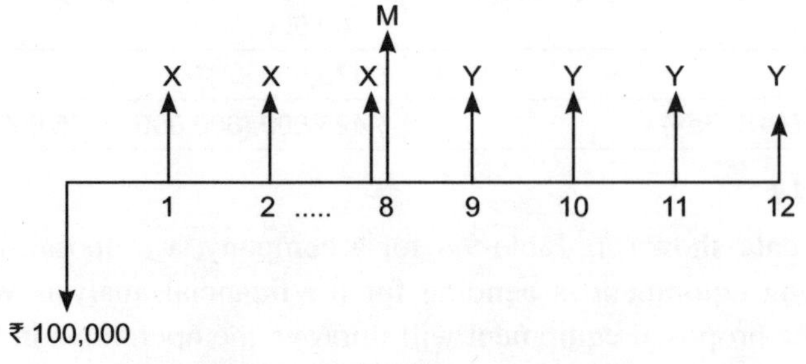

Fig. 9.11 Increased cash flow diagram

X = ₹ 30375,

And Y = 26000.

Now M = Value of annuity Y at the end of 8 years.

$$Y = \left[\frac{1-(1+r)^{-n}}{r}\right] = 26{,}000\left[\frac{1-(1.1)^{-4}}{0.1}\right],$$

because $n = 12 - 8 = 4$ years and $r = 10\%$, *i.e.* 0.1

therefore M = 82416.50

PV of Y from 9–12 years = PV of M at the end of 8 years

$$= \frac{M}{(1+r)^n} = \frac{82416.50}{(1.1)8} = ₹\ 38447.91.$$

PV of annuity X from 1–8 years

$$= X\left[\frac{1-(1+r)^{-n}}{r}\right]$$

$$= 30375\left[\frac{1-(1.1)^{-8}}{0.1}\right]$$

= ₹ 162048.38.

NPV = –100,000 + PV of annuity Y + PV of annuity X

= –100,000+ 38447.91 + 162048.38

= ₹ 100496.29.

Since the NPV is positive, the proposal should be accepted.

9.4.3 Profitability Index

Usually a company has many projects which await approval and thereafter implementation. The company will obviously like to undertake the projects which yield high profits per unit of investments. Net present value is a measure of profitability. NPV per unit of initial investment is called as profitability index.

$$\text{Profitability Index} = \frac{\text{NPV}}{\text{Intital Investment}}.$$

For example, if a firm has the following three proposed projects whose NPV and initial investments are as follows:

Projects	NPV (₹)	Initial Investment
X	30,000	35,000
Y	95,000	50,000
Z	1,30,000	1,10,000

From the data given above, profitability index for each project can be calculated. These are as follows:

Projects	Profitability index
X	0.86
Y	1.9
Z	1.18

Proposals should be arranged such that profitability indices are in decreasing order. This will be in the order of preference because higher profitability index is always desired.

Projects	Profitability index
Y	1.9
Z	1.18
X	0.86

Assume that the firm has a constraint on capital budget, *i.e.* it can invest only ₹ 60,000. In such a case, the firm should select project Y because it is on top priority based on profitability index.

9.5 Current Assets

Assets are mainly of two types:

(*a*) Fixed assets

(*b*) Current assets

Fixed assets include plant and equipment. These are also called fixed capital. Capital budgeting, which was discussed in previous section, is concerned with procurement decisions of fixed assets. Such assets are generally used throughout their life span.

On the other hand, current assets are mostly used in a short time period. Current assets include:

(*a*) Inventories

(*b*) Accounts receivable

(*c*) Cash as well as excess cash investment in short-term securities

Inventories include raw materials, purchased components, work-in-process inventory and finished good inventories. Raw material inventories will be formed into fabricated components, sub-assemblies and finally into finished goods. Finished goods inventory will be sold to consumers. Consumers pay for it, but usually there is delay in receiving the accounts from customers. In case of credit sales, this delay is even more. 'Accounts receivable' is a term used for such situations in which accounts are to be received from customers. Accounts receivable will finally be converted into cash.

Hence inventories if sold to customers yield into accounts receivable which will be converted into cash. The cash will again be used for purchase of raw materials. This is the cycle of current assets which is shown in Fig. 9.12.

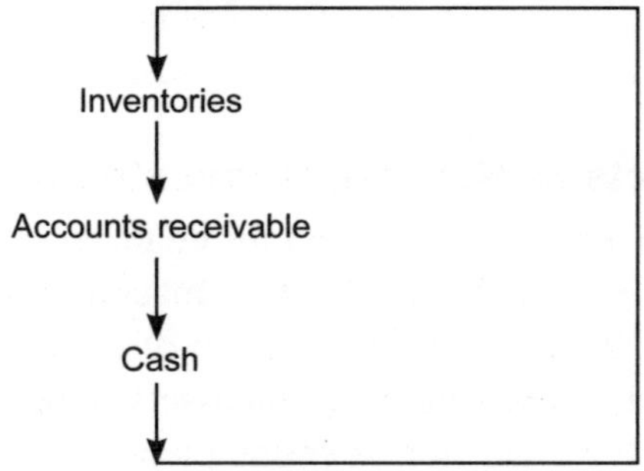

Fig. 9.12 Cycle of current assets

In Fig. 9.12, all the cash obtained is shown for purchase of raw material inventories for simplicity. However it is only one of the important use. In case of seasonal demands, the excess cash may be invested in short-term securities. Further, there are certain liabilities on the firm which are to be paid.

9.5.1 Net Working Capital

Current liabilities are associated with whatever is to be paid in the present accounting year. As the fixed capital is concerned with fixed assets, working capital is to current assets. Net working capital is defined as current assets minus current liabilities and it is a measure of liquidity.

Net working capital = Current assets – Current liabilities.

9.5.2 Aims and Objectives of Working Capital Policies

Working capital policies affect the management of current assets which include cash, inventories and accounts receivable, which in turn, affect the routine operations or day-to-day operations.

The day-to-day/routine operations include:

(*a*) Consumption of raw material

(*b*) Procurment of consumables pertaining to processes such as welding electrodes, furnace oil, etc.

(*c*) Wage payment of workers

(*d*) Consumption of resources including water, power, etc.

(*e*) Transportation of materials and goods

Day-to-day operations cannot be performed satisfactorily if sufficient cash is not available to the firm. To have sufficient money to spend on these activities is one of the main objectives of working capital policies.

As discussed before, the current assets are of circulating nature and proper level of current assets generates confidence in suppliers that their bills will be paid on time. One can view it as minimizing the risk.

Working capital policies should be aimed at maximizing the value of the firm. Effective management of current assets will indirectly contribute to increasing value of common shares because it provides a certain level of confidence in shareholders and. the company. This surely helps in maximizing the profit level of the firm which is the main objective of any business.

9.5.3 Current Assets as Reversible Investments

Net present value is an appropriate method for financial evaluation of fixed assets such as machines. This is because the life of the machine is estimated and that time period is used for calculating NPV. But in case of current assets, cash flow will occur up to the time till investment in current assets is there. As we stop investing in current assets, cash flow will stop after completion of cycle which is shown in Fig. 9.13. Here the time period, n, is irrelevant. In other words, as we withdraw the investments or reverse the investment, the cash flow will stop and therefore current assets are evaluated as reversible investments very often.

The policies concerning current assets are determined in such a way that either total annual cost is minimized or annual net profit is maximized.

Example 9.15

Show that the yearly net profit is independent of the time period when reversal of investment takes place

Consider the situation shown in Fig. 9.14.

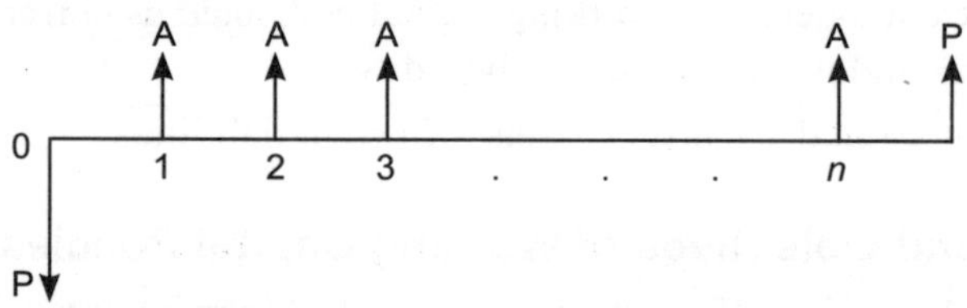

Investment of P is made at time period 0, *i.e.* at present and due to it annuity of cash flow A is obtained. It is decided to withdraw amount P will be withdrawn after n number of years and further cash flow will not be available. For simplicity, annuity is being considered. An analogy pertaining to investment in bank as all of us are very much familiar with it, can also be thought of. We invest amount P in bank and obtain yearly benefits in the form of interest. As soon as we withdraw the original investment, the benefits will stop flowing in.

Now NPV corresponding to profits obtained throughout n number of years = – P + P.V. of annuity A + P.V. of receipt of P at the end of n number of years

$$= \mathrm{P} + A\left[\frac{1-(1+r)^{-n}}{r}\right] + \frac{\mathrm{P}}{(1+r)^{n}}$$

Where r is a the discount rate.

$$\text{Now} \quad \text{NPV} = A\left[\frac{1-(1+r)^{-n}}{r}\right] + \mathrm{P}\left[1-(1-r)^{-n}\right]$$

$$= (A - \text{P}r)\left[\frac{1-(1+r)^{-n}}{r}\right] = \left[\frac{1-(1+r)^{-n}}{r}\right] \quad \text{... (9.13)}$$

Or $\quad \text{B} = A - \text{P}r. \quad$ ... (9.14)

We can also multiply the equation (9.13) with capital recovery factor, *i.e.* $\left[\frac{r}{1-(1+r)^{-n}}\right]$ in order to obtain the annuity corresponding to NPV for profits received throught n number of years and get the same result.

In Equation (9.14), n does not appear and therefore the yearly net profit is independent of the time period when reversal of investment takes places.

[**Note:** One can also understand it in another way. For example, if cost of capital is r, then at the end of the year 1, Pr will be paid to the investor and (A – Pr) is the net profit at the end of year 1. Similarly for subsequent years, (A – Pr) is the net profit and therefore the average yearly net profit is (A – Pr).]

9.5.4 Credit

It is a policy decision to offer credit to customers, *i.e.* allowing them to pay later. Customers may enjoy the services offered or receive the tangible products and then after some time, pay their dues. By extending the credit to customers, there is high probability of increasing the sales. But at the same time, a small percentage of the total consumers may not pay at all. This is called as bad debt loss and such losses are to be forecast or estimated.

Table 9.7 shows the comparison between situations before and after extension of credit to customers.

Table 9.7 Comparison between policies without and with credit facility

Without credit	With credit facility to customers
(1) Income I_1 = Periodic revenue – cash expenditure, *i.e.* variable cost.	(1) Income I_2 = Increased periodic revenue due to credit policy bad debt losses – variable cost
(2) Periodic income will start right from the first period	(2) Periodic income will start after credit period is over
(3) Cash flow diagram 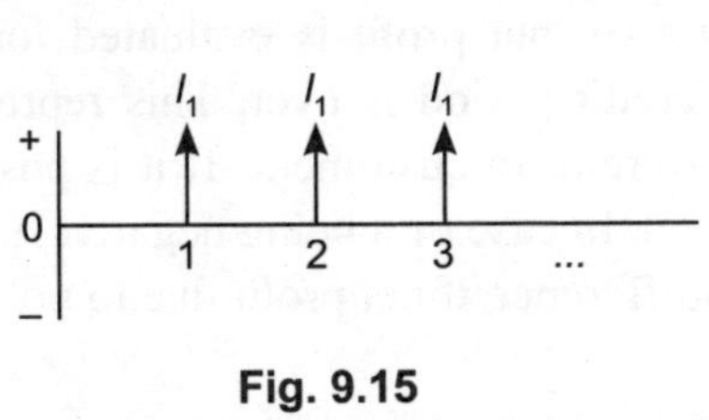 **Fig. 9.15**	(3) Cash flow diagram 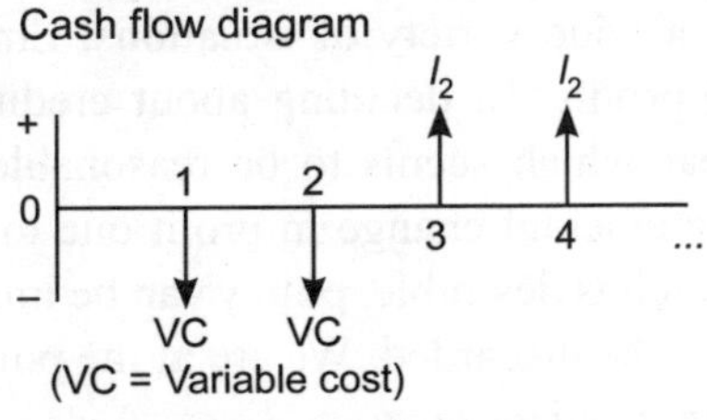 (VC = Variable cost) **Fig. 9.16**

Assuming credit for 2 periods where periods may be in months or weeks or even in days, the objective is to find out the net change due to credit policy which is shown in Fig. 9.17.

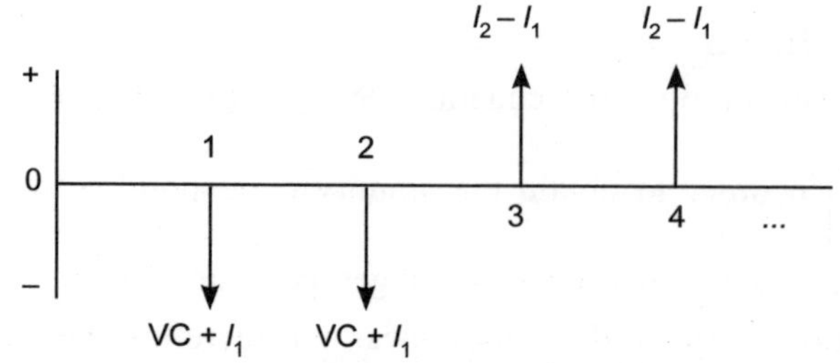

Fig. 9.17 Net change due to credit policy

As the variable cost is shown on the negative side in Fig. 9.16 and l_1 is on positive side in Fig. 9.15 for the credit period (assumed as 2 periods), net change in income is $(-\text{VC} - l_1)$ for first 2 periods, *i.e.* $-(\text{VC} + l_1)$ and therefore $(\text{VC} + l_1)$ is shown on negative side in Fig. 9.17. After the credit period, $(l_2 - l_1)$ is the net change in income per period shown in Fig. 9.17.

Annual change in income = $(l_2 - h)$ × No. of periods in a year (N)

(If periods are in months, N = l_2. l_1 in weeks, N = 520).

For the credit period there is no income. It can be considered as excess investment which is equal to $(\text{VC} + l_1) \times$ credit period and excess, capital cost is,

$(\text{VC} + l_1)$, × credit period × K,

Where K is cost of capital.

Assuming rate of tax as T,

Net profit = Annual change in income × $(l - \text{T})$ – excess capital cost

$= [(l_2 - l_1) \times \text{N} \times (l - \text{T})] - [(\text{V. C.} + l_1) \times \text{credit period} \times \text{K}]$.

It should be noted that variable cost corresponds to volume of sales. In the above equation for net profit, the VC corresponds to increased sales due to credit policy. Obviously if the net profit is positive, the credit policy can be recommended for implemention.

In the business environment, we are often facing the problems of decision making whether to implement particular policy or not. In all cases, more concern is towards incremental change in relevant parameters which are profit and cost in a wide variety of situations. Emphasis is on reduction of costs and increase in profits. In deciding about credit policy also, net profit is evaluated for one year which seems to be reasonable, after credit period is over. This represents incremental change in profit due to offer of credit to customers. If it is positive, which is desirable, policy can be implemented. In case of it being negative, policy may be discarded. We are at the point of indifference if net profit due to policy is evaluated as zero.

9.5.5 Average Collection Period

For example, if the monthly sales of a firm is ₹ 3 lakh and assuming 30 days in a month,

$$\text{Average sales per day} = \frac{300,000}{30} = ₹\ 10,000 \text{ per day.}$$

If the customers pay their accounts on an average in four days, then

Accounts receivable = 4 × 10,000 = ₹ 40,000.

In other words, one can define average collection period as accounts receivable divided by sales per period.

$$\text{Average collection period} = \frac{\text{Accounts receivable}}{\text{Credit sales/period}}.$$

9.5.6 Average Payment Period

The firm receives the money from its customers. But it has to pay to its suppliers also in connection with purchase of raw materials, components, etc. Average payment period is defined as accounts payable divided by purchases per day:

$$\text{Average payment period} = \frac{\text{Accounts payable}}{\text{Purchases per day}}$$

This section discusses the types of current assets. Working capital policies are also explained along with definitions of useful terms in connection with working capital.

EXERCISES

1. What do you understand by time value of money?
2. Explain the following factors:
   (*a*) Present value factor
   (*b*) Compound amount factor
   (*c*) Annuity present value factor
   (*d*) Capital recovery factor
   (*e*) Annuity compound amount factor
   (*f*) Sinking fund factor
3. An investment gives ₹ 90 as interest per year and ₹ 1000 at maturity at the end of five years from now. Assuming annual interest rate of 9%, evaluate the present value of this investment.
4. Differentiate between:
   (*a*) Shares and bonds
   (*b*) Secured bonds and unsecured bonds
   (*c*) Preference shares and equity shares

5. Explain the relationship between return and risk.

6. Present price of a share is ₹ 110. Probability corresponding to price after one year is as follows:

Price of the share after one year	Probability
80	0.05
90	0.09

❑❑❑

Bibliography

Chase R.B., Aquilano N.J., and Jacobs F.R. *Production and Operations Management*. TMH, New Delhi, 2000.

Nahmias S. *Production and Operations Analysis*. McGraw-Hill, 2001.

Panneerselvam R. *Production and Operations Management*. PHI, New Delhi, 1999.

Sharma S.C., Banga T.R., and Agarwal N.K. *Industrial Engineering and Management Science*. Khanna Publishers, Delhi, 2000.

Smith S.B. *Computer-based Production and Inventory Control*. Prentice-Hall, 1989.

Index

❑❑❑